Strategic Planning of Sustainable Urban Water Management

Strategic Planning of Sustainable Urban Water Management

Edited by Per-Arne Malmqvist,
Gerald Heinicke, Erik Kärrman,
Thor Axel Stenström and
Gilbert Svensson

Publishing

LONDON • SEATTLE

Published by IWA Publishing, Alliance House, 12 Caxton Street, London SW1H 0QS, UK
Telephone: +44 (0) 20 7654 5500; Fax: +44 (0) 20 7654 5555; Email: publications@iwap.co.uk
Web: www.iwapublishing.com

First published 2006
© 2006 IWA Publishing

Printed by Lightning Source.

Disclaimer

British Library Cataloguing in Publication Data
A CIP catalogue record for this book is available from the British Library

Library of Congress Cataloging- in-Publication Data
A catalog record for this book is available from the Library of Congress

ISBN13: 9781843391050

Contents

[v]

Contributors

EDITORS

Per-Arne Malmqvist, Urban Water Programme Director, Water Environment Technology, Chalmers University of Technology, Sweden. pam@urbanwater.chalmers.se

Gerald Heinicke, Ph.D., DHI Water and Environment AB, Lund, Sweden. gerald.heinicke@dhi.se

Erik Kärrman, Ph.D., Ecoloop AB, Stockholm, Sweden. erik.karrman@ecoloop.se

Thor Axel Stenström, Professor, Water and Environmental Microbiology, Swedish Institute for Infectious Disease Control, Solna, Sweden. thor-axel.stenstrom@smi.ki.se

Gilbert Svensson, Professor, DHI Water and Environment AB, Göteborg, Sweden and Dept. of Civil and Environmental Engineering, Luleå University of Technology, Sweden. gilbert.svensson@dhi.se

ADDITIONAL AUTHORS

Stefan Ahlman, M.Sc., Water Environment Technology, Chalmers University of Technology, Göteborg, Sweden. Email: stefan.ahlman@chalmers.se

Helena Almqvist, Ph.D., Municipality of Luleå, Sweden.
Email: helena.almqvist@tekn.lulea.se

Nicholas Ashbolt, Professor, Centre for Water and Waste Technology, University of New South Wales, Sydney, Australia. Email: n.ashbolt@unsw.edu.au

Andras Baky, M.Sc., JTI - Swedish Institute of Agricultural and Environmental Engineering, Uppsala, Sweden. Email: andras.baky@jti.slu.se

Anders Baun, Associate Professor, Institute of Environment & Resources, Technical University of Denmark, Lyngby, Denmark. Email: anb@er.dtu.dk

Jan-Olof Drangert, Associate Professor, Dept. of Water and Environmental Studies, The Tema Institute, University of Linköping, Sweden. Email: jandr@tema.liu.se

Eva Eriksson, Ph.D., Institute of Environment & Resources, Technical University of Denmark, Lyngby, Denmark. Email: eve@er.dtu.dk

Daniel Hellström, Ph.D., Stockholm Water AB, Sweden (water utility).
Email: daniel.hellstrom@stockholmvatten.se

Ulf Jeppsson, Associate Professor, Industrial Electrical Engineering and Automation, Lund University, Lund, Sweden. Email: ulf.jeppsson@iea.lth.se

Mats Johansson, M.Sc., Verna Ecology, Stockholm, Sweden. Email: mats@verna.se

Håkan Jönsson, Associate Professor, Dept. of Biometry and Engineering, Swedish Univ. of Agricultural Sciences, Uppsala, Sweden. Email: hakan.jonsson@bt.slu.se

Jaan-Henrik Kain, Ph.D., Dept. of Architecture, Chalmers University of Technology, Göteborg, Sweden. Email: kain@chalmers.se

Helena Krantz, Ph.D., Dept. of Water and Environmental Studies, The Tema Institute, University of Linköping, Sweden. Email: helkr@tema.liu.se

Anna Ledin, Professor, Institute of Environment & Resources, Technical University of Denmark, Lyngby, Denmark. Email: anl@er.dtu.dk

Marianne Löwgren, Professor, Dept. of Water and Environmental Studies, The Tema Institute, University of Linköping, Sweden. Email: marlo@tema.liu.se

Peter S. Mikkelsen, Associate Professor, Institute of Environment & Resources, Technical University of Denmark, Lyngby, Denmark. Email: psm@er.dtu.dk

Bo Olin, MBA, Sustainable Enterprise Solutions (Naturekonomihuset AB), Stockholm, Sweden. Email: bo@naturekonomi.se

Jakob Ottoson, Ph.D., Water and Environmental Microbiology, Swedish Institute for Infectious Disease Control, Solna, Sweden. Email: jakob.ottoson@smi.ki.se

Marika Palmér Rivera, M.Sc., WRS Uppsala AB, Uppsala, Sweden.
Email: marika.palmer-rivera@wrs.se

Susan R. Petterson, Ph.D., Centre for Water and Waste Technology, University of New South Wales, Sydney, Australia. Email: s.petterson@unsw.edu.au

Caroline Schönning, Ph.D., Swedish Institute for Infectious Disease Control, Solna, Sweden. Email: caroline.Schonning@smi.ki.se

Henriette Söderberg, Associate Professor, Dept. of Architecture, Chalmers University of Technology, Göteborg, Sweden, and Göteborg City.
Email: henriette.soderberg@cityhall.goteborg.se

Therese Westrell, Ph.D., Swedish National Food Administration, Uppsala, Sweden.
Email: therese.westrell@slv.se

Acknowledgements

This book is based on the results and experience from the research programme Sustainable Urban Water Management. The programme was financed by the Swedish Foundation for Strategic Environmental Research, MISTRA. Complementary financing was provided for some of the projects. Complementary financiers were, among others, VA-Forsk (the research foundation of the Swedish Water and Wastewater Association "Svenskt Vatten"), FORMAS (the Swedish Research Council for Environment, Agricultural Sciences and Spatial Planning), EU and others.

In the Urban Water model cities, the municipal water and wastewater organisations participated by taking active part in the applied research at their own expense. To the drinking water projects, Stockholm Water Co. (Stockholm Vatten) and Göteborg Water and Wastewater Works contributed considerable technical support, equipment, and chemical analyses.

Besides the editors and authors of the chapters in this book, a large number of researchers, consultants and municipal staff officials have made many kinds of contributions.

The Programme Board members have contributed significantly to the success of the programme and to the clear goal-orientation which has brought the research to a stage at which the results are actually in demand by municipal and regional organisations, and attract attention internationally. The members of the board are: Lars Gunnarsson, SYVAB (chairman); Peter Balmér, VA-strategi AB; Anders Berntell, SIWI; Carl-Johan Engström, City of Uppsala; and Birgitta Fritzdotter, VA-drift AB. They all have many years of experience with management in the water and wastewater sector, in municipal and regional organisations and in consultancy both in Sweden and abroad. Marie Uhrwing, the MISTRA officer for the programme, assisted the board.

The programme has had a reference group to give advice about management. The reference group included: Gunilla Brattberg, Stockholm Water Co.; Marie Larsson, the Swedish Environment Protection Agency; Jan Eksvärd, the Federation of Swedish Farmers (LRF); Inger Hansson, Karlshamns Water and Wastewater Works; Henrik Aspegren, Malmö Water and Wastewater Works; Kåre Olsson, the Swedish Society for Nature Conservation (SNF); Ulf Mohlander, the City of Stockholm Environment and Health Administration; Roger Bergström, the Swedish Water and Wastewater Association, "Svenskt Vatten"; Birgitta Olofsson, Tyréns Infrakonsult AB; Mia Thorpe, the housing cooperation HSB; and Bertil Gustavsson, City of Jönköping. Even more people participated in the reference group during the early stages of the programme.

The programme management group consisted of Per-Arne Malmqvist, Chalmers University of Technology and Luleå University of Technology; Thor Axel Stenström, the Swedish Institute for Infectious Disease Control (SMI), Gilbert Svensson, Chalmers University of Technology; Erik Kärrman, Ecoloop AB; and Henriette Söderberg, Chalmers University of Technology. During the first phase of the programme, Lars Bengtsson, Lund University, and Jan-Olof Drangert, University of Linköping, also participated in the management group.

Major project leaders and supervisors were, besides the researchers mentioned, Nicholas Ashbolt, University of New South Wales, Australia; Bengt Carlsson, Uppsala University; Gunnel Dalhammar, Royal Institute of Technology (KTH); Jörgen Hanaeus, Luleå University of Technology; Torsten Hedberg, Chalmers University of Technology; Daniel Hellström, Stockholm Water Co; Malte Hermansson, Göteborg University; Bengt Hultman, Royal Institute of Technology; Ulf Jeppsson, Lund University; Håkan Jönsson, Swedish University of Agricultural Sciences; Marianne Löwgren, Linköping University; Bo Olin, Sustainable Enterprise Solutions (Naturekonomihuset AB); and Helena Åberg, Göteborg University.

The Editors

Acronyms and Abbreviations

AHP	analytical hierarchy process
AKWA 2100	A research project in Germany (Alternatives of Municipal Water Supply and Sanitation)
AS	activated sludge
ASM 1	Activated Sludge Model 1 (also 2 & 3)
BMP	best management practices
BOD_7	biological oxygen demand, measured over 7 days
CBA	cost benefit analysis
CBR	case based reasoning
CE	cost effectiveness
CER	cost estimation relationship
COD	chemical oxygen demand
CRA	chemical risk assessment
CSO	combined sewer overflow
DCE	detailed cost estimation
DIM-SUM	Innovative Decision Making for Sustainable Management of Water in Developing Countries (EU project)
DMG	decision-making group

DOS	disk operating system
DS	dry solids
DSP	decision support process
EC	expanded clay
EMC	event mean concentration(s)
FBC	feature based costing
FORMAS	Swedish Research Council for Environment, Agricultural Sciences & Spatial Planning
GAC	granular activated carbon
GIS	geographical information systems
HACCP	hazard analysis and critical control point
HMP	hygiene modifying process
HRT	hydraulic residence time
IAP	ion activity product
IPPC	Integrated Pollution Prevention and Control
KREPRO	Kemwater recycling process. A process for extraction of phosphorous from sewage sludge.
LEO	local environmental office
MCA	multi-criteria analysis
MCDA	multi-criteria decision aid
MF	microfiltration
MIB	2-methylisoborneol
MISTRA	The Swedish Foundation for Strategic Environmental Research
MRA	microbial risk assessment
N & P	nitrogen and phosphorus
NAIADE	Novel Approach to Imprecise Assessment and Decision Environments (an MCDA)
NF	nanofiltration
NGO	non-governmental organisation
NN	neural network
NOM	natural organic matter
O & M	operation and maintenance
OCP	oxygen consumption potential
OECD	Organisation for Economic Co-operation and Development
ORWARE	Organic Waste Research
PAH	polycyclic aromatic hydrocarbons
PCR	polymerase chain reaction
PDF	probability density function
PE	parametric estimation
PEC	predicted environmental concentration
PKNS	phosphorus, potassium, nitrogen and sulphur

PNEC	predicted no effect concentration
PVC	polyvinyl chloride
QMRA	quantitative microbial risk assessment
REACH	EU legislation on chemicals
Rectangular rain	a method of describing a rain event
REGIME	an MCDA
RICH	ranking and identification of chemical hazards
RO	reverse osmosis
SCA	strategic choice approach
SEPA	Swedish Environmental Protection Agency (Naturvårdsverket)
SEWSYS	SEWer SYStem model
SFA	substance flow analysis
SIWI	Stockholm International Water Institute
SMC	site mean concentration
SRT	solids retention time
STP	sewage treatment plant
STRAD	Strategic Advisor
SWARD	Sustainable Water Industry Asset Resource Decisions
TDI	tolerable daily intake
TGD	Technical Guidance Document in support of a Commission Directive (EU)
TS	total solids
TSS	total suspended solids
UASB	upflow anaerobic sludge blanket (reactor)
UF	ultrafiltration
URWARE	Urban Water Research (model)
VA-Forsk	research foundation of the Swedish Water & Wastewater Association "Svensk Vatten"
WFD	Water Framework Directive (EU)
VS	volatile solids
WTA	willingness to accept
WTP	willingness to pay
ww	wastewater
WWTP	wastewater treatment plant
XOC	xenobiotic organic compounds

1

Urban water in context

Per-Arne Malmqvist, Gerald Heinicke, Erik Kärrman,
Thor Axel Stenström and Gilbert Svensson

1.1 THE CHALLENGE

In industrialised countries, water and sanitation is an integrated part of the urban planning and service to citizens. In most cases, the sector is conservative and characterised by *business as usual*. Further development is initiated by the connection of newly built or peri-urban areas to the existing systems, or by minor adjustments, such as the addition of new treatment steps, made necessary by new regulations or emerging problems discovered in sector-limited investigations. However, system analytical questions are of growing concern.

Urban water systems in Western Europe and North America seldom suffer from an acute crisis related to water provision and basic management including operation and maintenance. Nevertheless, there is an imbalance between the obvious needs for improvement and the actual implementation, with respect to water infrastructure rehabilitation, upgraded wastewater treatment, and the integration of urban water management with ecological requirements, which include the recycling of nutrients from wastewater and a decrease of diffuse pollution. In developing countries, one of the greatest challenges of population growth and urbanisation is the supply of water and sanitation services. According to a WHO and UNICEF report (2000), 2.4 billion people lacked access to improved sanitation at the beginning of 2000. The Millennium Development Goals strongly emphasise two specific goals related to water:

- To halve by the year 2015 the proportion of people who are unable to reach or to afford safe drinking water; and
- To stop the unsustainable exploitation of water resources by developing water management strategies, at the regional, national and local levels, which promote both equitable access and adequate supplies.

In addition to these goals, the proportion of people lacking access to adequate sanitation should be halved by 2015. Sanitation systems must be designed to safeguard human health as well as the health of the environment.

The urban water infrastructure is characterised by long lifetimes and huge investment costs. Once built, there is little freedom to opt for changes, for technical, organisational and financial reasons. The choices made now will affect the life quality of many people, society's use of resources, and our environment. The interrelations to be considered extend far into the future, with a corresponding uncertainty of information, regarding, among other things, migration, climatic change and technological development. It is therefore essential to have a good quality strategic planning! This task requires flexible and reliable planning tools, to predict forthcoming problems and to find suitable management approaches. The predictions may be about microbial or chemical risks, as well as the environmental impact, of hitherto hypothetical system alternatives. The comprehensive approach needed should emphasise not only the technical and economic aspects but also the challenges of institutional capacity and public participation in the planning process. There is also a need to integrate the urban area into the whole catchment area, to facilitate accurate analyses and avoid sub-optimisation. A "toolbox" for finding answers to these questions is needed, and also case studies in which the tools have been tested and validated.

In addition to the Urban Water Management program in Sweden, two other major European research programmes were started, in the late 1990s, with similar starting positions and problems concerning the design of future water and wastewater solutions. They both focus on the strategic development of urban water systems, covering technical and organisational aspects based on the whole chain from water supply to wastewater handling.

In Germany, the AKWA 2100 project (*Alternativen der kommunalen Wasserversorgung und Abwasserentsorgung*, Alternatives of municipal water supply and sanitation) investigated three alternative strategies, for developing the urban water infrastructure, and compared two source-separated systems with an improved conventional system. One of the source-separating systems was designed with pipe-bound collection of urine for nutrient recovery. The other one was a decentralised system with a self-sustaining water supply as well as water treatment and recycling, within each building. All three systems were hypothetically applied to two municipalities, one urban and the other peri-urban. The assessments included economy (with a time frame of 150 years) and aspects of sustainability. The analytical hierarchy process (AHP) was applied as a multi-criteria decision aid. The results indicate that, in the long run, the decentralised, high-tech source-separating systems are not unreasonably more expensive than the conventional one, especially when implemented from the start in a newly built area. Moreover, decentralised source separation was superior to the other systems in the other categories of sustainability also, *i.e.* the environmental and social aspects. Centralised urine separation was found to offer only a small advantage over an improved conventional system; it was also the most costly alternative. The project demonstrated that alternative water and wastewater systems may be both resource efficient and economically sound, as well as that the transition is technically feasible (Hiessl and Toussaint 2003; Hiessl *et al.* 2003). One of the conclusions was that, although the technology for an in-house water and wastewater cycle may be available, its application does not appear to be well enough established for city planners to take it into account at present. Furthermore, tight finances may force decision makers to opt for solutions that are inexpensive in the short term.

In Britain, the Sustainable Water Industry Asset Resource Decisions (SWARD) procedure was developed to facilitate the inclusion of sustainability issues in decision-making processes in a structured way. In their procedural guide for sustainable water services (Ashley *et al.* 2004), the authors put forward a series of seven successive decision support processes (DSPs). These cover the whole chain from defining objectives to post-project monitoring. The criteria include economic, environmental, social and technical aspects, which are substantiated by corresponding indicators. The procedure for decision support was then applied in two case studies, one on water supply and the other on strategies to manage domestic sanitary waste disposed via the WC (Butler *et al.* 2003; Ashley *et al.* 2004).

In comparison, the Urban Water programme is more diverse. The projects included the development of tools described in this book, as well as additional research projects in related fields. Integration of the research with the water and wastewater sector has been on a municipal level in numerous projects, with a wide geographical and thematic spread. Moreover, a significant impact is expected in that new experts, experienced in strategic planning of urban water systems, will become available to the industry, to municipal planning departments, and to research institutes.

Will the sustainable water and wastewater systems of the future be improved versions of what exists today, or will there be some radical changes?

The answer to this question cannot possibly be a single one that would cover all planning situations. On the contrary, each city has to make strategic decisions based on the local context. Although the Urban Water programme does not give answers, it offers decision support to the planners of future sustainable water and wastewater systems. This book offers the reader inspiration, some decision support tools that are available for systems analysis and planning of urban water systems, and practical examples of their application. Some of the driving forces for change are universal; others are local and specific. The tools presented are, or may be, adapted to the circumstances present in both developed and developing countries.

1.2 URBAN WATER IN A NUTSHELL

The Urban Water programme (www.urbanwater.org) has been a long-term integrated research programme including senior researchers and PhD students from 9 Swedish universities. It was initiated as a response to the general demands of the national water and wastewater sector. The decision-makers' uncertainty was due not only to their awareness that the renewal and upgrade of water systems will require large investments in the coming decades but also to questions whether the current systems have the potential to be embrace ecologically sustainable solutions.

A conceptual framework has guided the projects in the Urban Water programme. A meaningful analysis requires that the definition of an urban water management system include the technical structure, the organisation, and the users. A comprehensive analysis of sustainability should cover five perspectives: health, environment, economy, socio-culture and technical function (Figure 1-1).

Each perspective is linked to both physical and non-physical criteria for sustainability. Physical criteria are impacts on the environment (environmental criteria) and society (health & hygiene criteria), while non-physical criteria are socio-culture, economy, and to some extent technical function. Each group of criteria contains at least one indicator or a set of indicators for the assessment of sustainability. An indicator is a variable that can be given a quantitative or qualitative value. Most of the methods aim to quantify the indicators, some of which are based on thresholds and involve specific requirements that have to be met. Others are relative and are intended for comparisons of two or more alternatives.

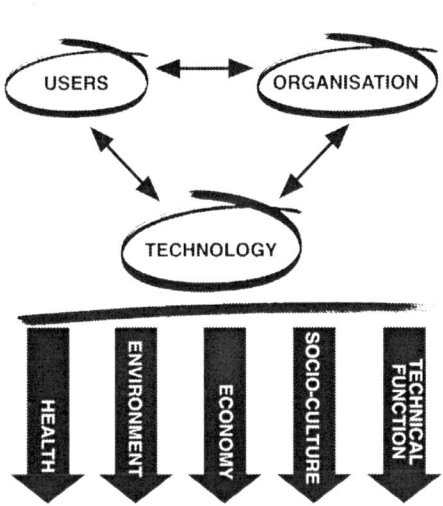

Figure 1-1. The Urban Water conceptual framework.

The lists of criteria and indicators vary for differing applications, depending on the local situation, the priorities of the stakeholders, the driving forces, and the problems that have been recognised. The principles for their selection are formulated for each application. One of the most important principles, or restrictions, applies to the availability of information. Experience has shown that the final number of criteria should be kept low, since this facilitates efficient work and enables multi-criteria decision making with involvement of stakeholders. An example of such a list is given in Table 1-1. This example concerns strategic planning for sustainable wastewater management in the City of Uppsala. The set of criteria and indicators was developed in co-operation between a researchers group and employees from various departments of Uppsala municipality, and then applied in a decision-making workshop (Section 9.5).

Table 1-1. Criteria and indicators used for systems analysis of the wastewater system in the Uppsala study (Section 9.5).

Criterion	Indicator
Health and hygiene criteria:	
Risk of infection	Number of infections per year
	Number of disease cases per year
Environmental criteria:	
Eutrophication	P to water (t/year)
	The share of P flows from wastewater systems to the recipient (%)
Nutrient recycling	N, P, K, S to agriculture (t/ year)
	N, P, K, S to other applications (t/year)
Toxic compounds to soil	Cd to agriculture (kg/year)
	Cd to other soils (kg/year)
Energy use	Energy use (kWh/person, year)
	Energy recovery (kWh/person, year)
Socio-cultural criteria:	
Value coalition between crucial actors	Qualitative
Action space: legislative political support	Qualitative
Presence of policy entrepreneurs	Qualitative
Confidence in the environmental benefits of a system	Qualitative
Trust in actors involved	Qualitative
Economic criteria:	
Annual cost	Annual cost for investment, operation and maintenance (€/year)
Functional and technical criteria:	
Technical function	Accidental risks (qualitative)

A prerequisite for the implementation of many of the criteria is an institutional system in place, which includes:

- Legislation that allows for sustainable solutions;
- A political system that allows for a fair and open decision-making process;
- Trustworthy institutions; and
- Key stakeholders who are in favour of improvements, such as institutional and technical changes.

1.2.1 The urban water toolbox

The criteria and indicators have been linked to the development of some strategic tools within the Urban Water research programme. The tools are intended to facilitate a structured analysis and an assessment of selected aspects of urban water systems; they also enable researchers and planners to conduct comprehensive and multidisciplinary analysis as a good basis for decision-making. The results of the analyses in different fields (*e.g.* environmental and economic), and their ranking of alternatives, need to be accounted for in methods to support decision-makers by helping them to synthesise and integrate the results (Section 8). Some of these tools are briefly summarised below.

Substance flow analysis

For many types of assessment, it is necessary to know how water and selected substances are moving through the technical systems, and where the latter eventually remain: in the receiving water, the sewage sludge, or in the air. For effective countermeasures, it is also necessary to know the origin of hazardous substances. Two tools for substance flow analysis have been developed and tested: URWARE and SEWSYS.

The URWARE (Urban Water Research Model) is a substance flow model that builds on Matlab and Simulink software. Its purpose is to simulate flows of various substances in wastewater systems. The model allows for the calculation of emissions to air, water and soil, as well as energy consumption. URWARE is a "library" of mathematical models that cover many aspects, including the activities in households, drinking water, transports, wastewater treatment, sludge treatment, incineration, and waste disposal. The model can be adapted to a specific sewage system being investigated, with either location-specific input data or default values that are included in the program. Environmental impacts that are normally covered by LCA analyses can be calculated by URWARE. These are global impacts, such as the greenhouse effect and acidification, and local impacts, such as eutrophication (Section 2.1).

The SEWSYS (Sewer System Model), which also uses the Matlab/Simulink platform, is a material flow model for the analysis of stormwater systems and combined sewers. Several types of stormwater treatment, such as ponds and infiltration, have been built into the model. In its present form, SEWSYS can analyse 20 variables (substances and combined parameters) from their sources to the discharge into a recipient. The sources include households, buildings, and traffic areas. Input data may be acquired from measurements or maps, or downloaded directly from a geographical information system (GIS). The recipients include soil, water, air, and sewage sludge (Section 2.2).

Microbial risk assessment (MRA)

The MRA methodology (Section 3.1) relates to the criterion risk of infection and comprises the classic steps of risk assessment: to determine hazards and exposure, to find dose-response correlations, and to make an assessment of the risk. The method is based on the identification of the times, frequency and places (exposure points) at which human beings are exposed to pathogens from wastewater, greywater, and drinking water within the defined system boundaries of the study. For each exposure point, the exposure and the risk are then quantified by conducting a predictive simulation that also accounts for variability. The barrier efficiency of water treatment processes is integrated in the model. The risk assessment can be performed for pathogens from all the relevant groups: bacteria, viruses, protozoa and worms. The method is in accordance with the guidelines published by the World Health Organization (WHO 2004).

Chemical risk assessment (CRA)

The chemical risk assessment (CRA) tool can be used to identify the substances in stormwater and wastewater that pose major risks to human health or the environment. The first step is to identify the critical elements, in terms of their toxicity, bioaccumulation and persistence characteristics. Then a site-specific study of the concentrations of these substances in water and soil is made; the levels found are then compared with limit values established by authorities, or with limits determined in relation to injury to living organisms. On the basis of the knowledge gained, measures are proposed to reduce risks, based on the kinds of "barriers" that are relevant for the risk substances identified. These barriers may include making laws and regulations, the choice and design of the technical system, treatment processes, and increasing public awareness (Sections 3.2 and 3.3).

Economic factors

Business economic costs are estimated with a calculation tool, developed in Excel, which includes investment as well as operation and maintenance. When alternative components within a larger system are assessed, both a "new-establishment" calculation and a "renewal" calculation are made. Either default data, site-specific information, or both can be used. Environmental and societal costs are, in addition to the straightforward expenses, included in a "value-added" model, which calculates the benefits of a wastewater management system, such as nutrient and heat recovery, as well as recipient quality (Section 4.1).

Organisation

What essential qualities are needed for an organisation to own, plan and operate a sustainable wastewater management system? Under what conditions can visions and plans be implemented in practice, and when is implementation impossible? At Urban Water, practical guidelines have been developed for decision-makers, based on criteria to consider when making comparative assessments between alternatives (Section 5.1).

Households

The Urban Water research has identified crucial factors pertaining to households and users. For example, to what extent are households capable of contributing to financing and maintenance? The experience has been transformed into practical guidelines for the planning and structuring of urban water systems. The recommendations provide support for local organisations in planning and operating systems that are based on new or alternative technology (see Section 5.2).

1.2.2 Using the tools

The above-mentioned tools have been tested in several Swedish urban areas, further exemplified in the following chapters. The investigations are based on specific questions, for example the following.

- In a newly built area (Hammarby Sjöstad, Stockholm), the local politicians expressed the ambition to develop an environmentally friendly wastewater system. A strategic planning process was initiated (Section 9.1).
- A small town (Surahammar), uses food waste disposers in households. Integrated food waste and wastewater handling versus separate systems were analysed (Section 9.2).
- In an urban enclave with ecological housing (Gebers in Stockholm), the use of urine separation and dry toilets was evaluated (Section 9.3).
- In a city centre neighbourhood, Vasastan in Göteborg (Gothenburg), the problem of combined sewers was studied (Section 9.4).
- In a suburban area built in the 1960s (Gottsunda, Uppsala), a question was raised about whether sorting of wastewater would increase the capacity of the existing central system.
- For the city of Uppsala, the experiences from earlier model city work were synthesised. Alternatives for wastewater and stormwater

handling were analysed, and a strategic planning process was tested
for wastewater management for the year 2020 (Section 9.5).

In parallel, but apart from the research programme, some of the methods
and tools were applied in other situations. For a peri-urban area (in
Södertälje), the municipality and the residents were assisted in the choice
between connection to the central system or local wastewater treatment
facilities. Some northern cities (Sundsvall and others) investigated solutions to
the problem of sustainable snow handling, such as the choice between central
snow deposits, local deposits near buildings, or dumping the snow into the
sea.

Methods from the toolbox are currently being applied to devise strategies
for the future wastewater handling in Sweden's two largest cities, Göteborg
and Stockholm. The studies also cover the problems of how to prevent
polluted stormwater (with a high content of *e.g.* heavy metals) from
contaminating the sewage sludge or the recipient. Internationally, the tools are
being applied in the EU project *DIM-SUM*, which aims to develop sustainable
water and wastewater systems in India, Malaysia, Indonesia, and Nepal.

1.3 URBAN WATER IN PRACTICE

1.3.1 Strategic questions

The methodologies and tools described in this book may be used for the study
of a variety of problems and situations. Some of the strategic questions dealt
with are the following.

Should water and sanitation be decentralised in peri-urban
        areas?

Connecting peripheral residential areas to the central water and wastewater
management systems often involves great expense to the residents. As an
alternative, a local solution is often less expensive. This may involve small
wastewater treatment facilities, such as the use of infiltration and wetlands, or
a type of source separating system. However, the funding and distribution of
responsibility may be complicated, particularly in relation to operational
disruptions and maintenance, and the subsequent impact on humans or the
environment. Here, the planning process is a critical link.

Wastewater systems: sorting or conventional?

Under certain circumstances, source separating wastewater systems can be
more sustainable than conventional ones. In Sweden, various sorting systems

have been introduced and tested during the past decade. Such systems include the separation of urine and, after storage, direct use of it in agriculture. Another option is the separation and treatment of blackwater, for the same purpose.

In a sustainable society, the use of plant nutrients contained in wastewater needs to be addressed. Sorting systems make it possible to recycle some of these nutrients to arable land. This may also be possible in a central wastewater system, for example by the direct use of sludge or wastewater, after hygienisation, or by the extraction of phosphorus from wastewater, sludge or ashes. The choice of strategy is highly dependent on factors such as the quality of the sludge, transport distances and storage options, cost of investments and energy, the willingness of farmers to use the product, and the readiness of the municipal administrations to implement the system change.

Will new drinking water treatment be more secure or sustainable?

New, flexible drinking water treatment processes have been developed and introduced during the past decades. With the development of modern technologies, such as membrane- and biofiltration, as well as alternatives to chlorination, it has become possible to choose the scale for water treatment and distribution. Local water treatment facilities would allow shorter distribution systems and thereby help improve the quality of the tap water.

Are open stormwater solutions an option?

Stormwater is a problem in many cities, causing flooding, overflows in combined sewer systems, and the pollution of sewage sludge with heavy metals and hazardous organic compounds. In other respects, stormwater is essential to urban areas, where it recharges both the groundwater and the watercourses. Open stormwater solutions, such as ponds, wetlands and green roofs, have become increasingly interesting, both as potential responses to technical or environmental problems, and as adding desirable elements to the urban landscape. The Urban Water toolbox helps you to find the preferable solution.

Can compounds in sewage sludge be beneficially used?

In the future, various treatment methods would allow for the use of sewage sludge as raw material for fertiliser products, energy production, and construction materials. Treatment plants may have a separate sludge fractionation step, with an extraction of plant nutrients. Alternatively, biological techniques may be used to extract products before the digestion chamber, after which the sludge is treated by supercritical water oxidation. The Urban Water toolbox offers methods to investigate the feasibility of such techniques.

How can the recycling of solid organic waste be integrated?

When designing a system for the management of organic household waste, *e.g.* choosing between central composting and food waste disposal units in the kitchens, the environmental and ecocyclic aspects are crucial, as are the economic aspects. From a technical point of view, disposal units that add organic waste to the sewer may be problematic due to blocked pipelines both at the property and at the municipal wastewater network. In general, food waste disposal systems have been found to be more expensive than central composting for households; however there are exceptions. The Urban Water toolbox helps you compare the system alternatives available.

1.3.2 Strategic planning

Strategic planning is an integrated, comprehensive approach that emphases not only the technical and economic aspects, but also the challenges of institutional capacity and public participation, in the process of sustainable water and wastewater management. The Urban Water programme has developed a comprehensive decision support tool, which includes sustainability criteria, knowledge of various technical solutions, and generic methods for information evaluation and presentation.

All planning is situation driven and must be carried out in the local context. Leading questions are:

1. What are the problems to be solved?
2. What are the driving forces for future development? What can be expected in the future?
3. Who are the stakeholders, *i.e.* who should be included in the planning process?
4. What alternatives do we have?
5. How shall the knowledge, decision-making and commitment of the stakeholders involved be integrated?

The planning process is iterative; it may be repeated until an acceptable consensus is achieved. A further component is a comprehensive systems analysis of the strategic alternatives selected. In many countries, the planning process is regulated such that the planning administrations must consult the public. It is essential to involve all major stakeholders early in a project.

Planning processes have failed or been delayed, when overlooked but influential stakeholders became involved late in the process and then blocked the decisions, even those that were motivated and rational.

Participation in the planning process

The participation in a planning process may be based on reasons ranging from manipulation (*i.e.* for easier implementation), to knowledge gathering (*i.e.* they know something that we need to know), to real decentralisation of influence for those who want to share power. Some questions should be addressed at the beginning of a study, such as:

- *Why* are we interested in participation? The answer influences the whole planning process.
- *Who* is to participate?
- *When* is stakeholder participation to take place?
- *How* will participation be facilitated?

Ideally, all users and stakeholders of water and wastewater services are involved, including households, industry and others. Stakeholders are also the organisations and authorities that own, plan, operate, finance, issue permits or control the water services, as well as housing companies and farmers. Participation of users in the planning process is crucial, especially for local systems, and it brings the water services closer to the people involved.

Although there are several models for stakeholder involvement (*e.g.* citizen advisory committees, planning cells, citizen juries, study groups, focus groups, expert panels, policy exercises and others), they are often employed as appendices to the processes of knowledge integration and decision-making. Stakeholders may contribute more to knowledge integration, as a process of individual and organisational learning, if they are involved more intensely in the planning and decision-making process as a whole (Argyris and Schön 1996; Friend and Hickling 2005).

Defining problems, options and potential strategies

A suitable guiding framework for the process of choosing a sustainable water and wastewater system is the strategic choice approach (SCA, see Friend and Hickling 2005), which is thoroughly tested in decision-making and management of uncertainties. The SCA process comprises four interconnected modes, linked to each other in an iterative and participative process.

1.   Strategically relevant questions are selected to shape the *problem focus*.
2.   The options are identified, after which potential strategies for addressing the problems are *designed*.
3.   To evaluate and *compare* the potential strategies,
      a) criteria and indicators for assessing the potential strategies are selected,
      b) the information needed is introduced into the process as preparation, and
      c) the process of knowledge integration is begun.
4.   While *choosing* among the potential strategies, the process moves towards a phase of decision-making and building of commitment among stakeholders. Lack of knowledge and remaining uncertainties are also identified.

Based on future scenarios, SCA facilitates linking problems with options, for example:

- Water shortage, to which demand management and wastewater management can respond;
- Willingness to pay, which can be increased by fair consumption fees, an efficient organisation and high reliability of the services;
- Groundwater pollution, which can be reduced by improved wastewater management.

When a suitable decision scheme has been defined, the process of knowledge integration may enter the evaluation phase.

Thematic and systemic knowledge

Information may be quantitative or qualitative, written or illustrated, precise or ambiguous, and established or debatable. Components in the acquisition and evaluation of information are:

- Basic information on the local situation (baseline study);
- Future scenarios, driving forces and pressures, and priorities;
- Service providers' performance and service levels;
- Screening of technological options; acceptance of options by users and the organisations involved;
- Mapping of institutional frameworks, their potential, adaptability and flexibility;
- Inventory of perceived, as well as quantified, risks;

- Collection of thematic information (socio-economy, environmental assessment, system risks and system robustness data);
- Adapting information to forms that are suitable for multi-stakeholder processes;
- Formulating criteria and indicators that are specific for the local conditions;
- Integration of information and building of commitment among stakeholders; and
- If necessary, methods for conflict resolution.

Evaluation of potential strategies

To evaluate potential strategies, the first step is that the participating stakeholders define relevant criteria and indicators. The types of criteria and indicators will probably be heterogeneous. If so, the formulation of criteria, preferences and indicators needs to be supported by methods that facilitate deliberative processes. The second step deals with how the diverse kinds of information may be combined as a base for the strategic assessments. Thematic, model-based and systemic types of knowledge are brought into the process. Here, the strength of the SCA method lies in managing manifold and diverse types of information by incorporating both qualitative and quantitative input, as well as scientific and lay knowledge. This is supported by multi-criteria decision aids. Established multi-criteria support tools, such as Naiade, STRAD or Regime, may be used (Section 8).

Decision-making and commitment

The last phase of the strategic choice approach (SCA) cycle, the *choosing*, has two aspects: the management of uncertainty and the process of decision-making. Uncertainty is central to the procedure: decisions are made step by step, while some uncertainty remains. As the uncertainties are defined, options to explore them are identified. The aim of the procedure is to reduce a wide range of potential strategies to a limited number of principal strategies for decision-making.

The SCA focuses on the timing of decisions: immediate actions, deferred choices (*i.e.* decisions that need to be taken in the future), and contingency planning. When methodological support is needed for conflict resolution, drama theory and confrontation analysis, for example, may assist in mapping power structures.

1.3.3 Systems analysis in practice: the Göteborg case

An ongoing project in Göteborg, a city with about 600,000 inhabitants, aims to develop a strategy for the future wastewater handling up to the year 2050. The existing conventional system has a single large wastewater treatment plant (WWTP), which receives the wastewater from the whole region. The WWTP is modern and meets all current and anticipated requirements for biochemical oxygen demand (BOD), phosphorus and nitrogen. The sludge is used in the production of soil for gardening in public areas, such as parks and roadsides, and for covering solid waste deposits. The long-term ambition, however, is to reuse the nutrients contained in the sludge for agriculture.

Stakeholders and organisation

The administration driving the project forward is the Department for Sustainable Water and Waste, a relatively young organisation which has been given the responsibilities of administering water, wastewater and solid waste handling in the city; this includes strategic planning and consumer contacts. The department orders services from, among others, the Göteborg Water and Wastewater Works, which is responsible for the production and distribution of drinking water and also for the collection of wastewater and stormwater. The regional *Gryaab* Company is responsible for wastewater treatment and the transport of wastewater to the WWTP by means of an extensive bedrock tunnel system. In 2004, the Department for Sustainable Water and Waste was given the task, by a political board, to investigate the future strategy for wastewater handling. A project board was formed, which includes representatives from the Department for Sustainable Water and Waste, the Water and Wastewater Works, and Gryaab. The project group, responsible for conducting the project, consists of experts from the three organisations, and researchers from the Urban Water programme.

Although it was realised that many more organisations and people would have a stake in the matter, it was decided that these would be consulted on specific occasions, instead of being involved in the direct project work.

Project plan

An approach for the Göteborg project, according to the strategic choice approach (SCA), was selected and structured in the following phases:

1.   Formulating the problem,
2.   Designing strategic options,
3.   Comparing these options, and
4.   Choosing a strategy.

At present, the project is in the phase of *comparing options*.

Driving forces

The assignment given to the Department for Sustainable Water and Waste included planning for as much recycling of nutrients from the wastewater and solid waste systems as is feasible in the long term. This, in turn, stems from the Swedish Environmental Goals, adopted by the national parliament. A current law proposal requires a 60% recovery of the phosphorus contained in wastewater, and its use in agriculture. Recycling ambitions also include other resources often contained in commercial fertilisers, such as nitrogen, potassium and sulphur, but without quantitative goals for these elements.

Scenarios

Scenarios are here defined as possible future outcomes. Extrapolations to 2050 of current trends were discarded. In a stakeholder workshop, the two most important factors concerning the choice and performance of the wastewater system for the year 2050 were identified: 1) the degree of sustainability achieved by society, and 2) the level of economic and social strength (economic wealth) in the region. Alternatives were described in a quadrant-type diagram with two future outcomes chosen for the screening phase. One is a target scenario; the other is a worst-case scenario. In Table 1-2, which summarises the assessments, the worst-case scenario may also be seen as a kind of sensitivity analysis for the first scenario. In reality, only some of the threats are expected to occur at the same time.

Table 1-2. Scenarios in the Göteborg systems analysis project.

Parameter	2050 Target scenario	2050 Worst-case scenario
Price for energy	Twice today's price	10 times today's price
Price for phosphorus	Twice today's price	10 times today's price
Use of hazardous chemicals	Half of today's use	Same as today
Impact of medicine residues	Half of today's impact	Same as today
Labour cost	Same as today	Five times today's cost
Restrictions		
- On transports	Yes, some	Yes, far-reaching
- On the use of energy	Yes, some	Yes, far-reaching

Criteria for sustainability

A short list of the criteria was chosen for the Göteborg project:

- Recycling of phosphorus to cultivated land;
- Protection of receiving waters;

- Protection of soil, *i.e.* no accumulation of hazardous compounds;
- Protection of human health, by reduction of toxic compounds;
- Energy use, including transports;
- Total cost of treatment;
- Total costs of transports; and
- Technical function.

The recycling of nitrogen is also a goal, since it is probably the nutrient most appreciated by the farmers. However, the feasibility of nitrogen recycling has been regarded as a question of energy only, because nitrogen is a renewable resource and can be obtained by fixation from the air. In the final evaluation, a more comprehensive list of criteria will be used and complemented by hygienic and social criteria.

Designing strategic options

A large number of potential alternatives were explored and briefly evaluated according to common knowledge and the local circumstances. The result was a short list of feasible options. Among the choices discarded were systems that require large areas, *e.g.* constructed wetlands. In a screening procedure, the project group graded the system alternatives qualitatively using a 5-degree scale. The results were discussed and agreed upon by both the project group and project board, which also excluded alternatives requiring extensive use of transports by lorry. The alternatives selected for further investigation and comparison are two types: centralised systems and those with source separation of wastewater fractions.

Centralised systems:
- 1a. Today's wastewater system, but with sludge incineration and landfilling of the ashes.
- 1b. Today's wastewater system and sludge handling (Figure 1-2)
- 1c. Today's wastewater system, plus far-reaching control of diffuse pollution in society, *e.g.* by physical separation of the combined sewers, and by replacement of in-house copper pipes in favour of stainless steel or polyethylene.
- 1d. Today's wastewater system, plus extraction of a pure phosphorus product from the water phase, *e.g.* by advanced biological processes.
- 1e. Today's wastewater system, plus extraction of a pure stream of several nutrients from the water phase, *e.g.* by membrane processes.

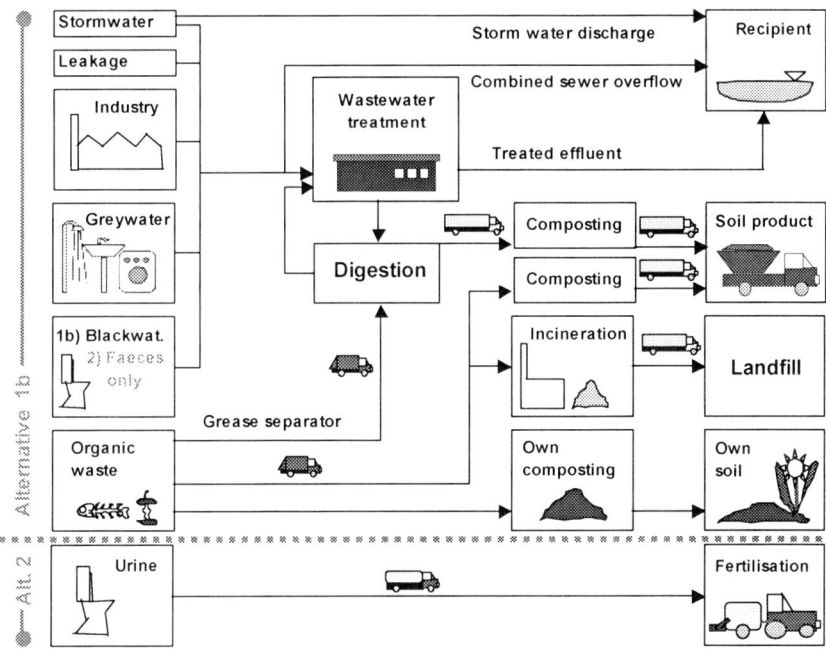

Figure 1-2. System alternatives 1b) Today's system, and 2) Urine separation.

Systems with source-separation of wastewater fractions:

2. Urine separation in the households, followed by lorry transport directly to storage facilities near a farm, and subsequent spreading on arable land (Figure 1-2).

3. Blackwater (WC) separation with pipe transport to one of several semi-local treatment plants (Figure 1-3).

Additional far-reaching source-separating alternatives will be investigated qualitatively. These include 4) dry toilets and 5) a completely local system, comparable to the one described in the AKWA study (Hiessl *et al.* 2003), with drinking water supplied by treated rainwater, and integrated wastewater treatment and reuse in each house or block of flats. Some of the alternatives will be combined with food waste disposers, and others with measures for the improvement of stormwater quality in the parts of the system that will also have combined sewers in the future.

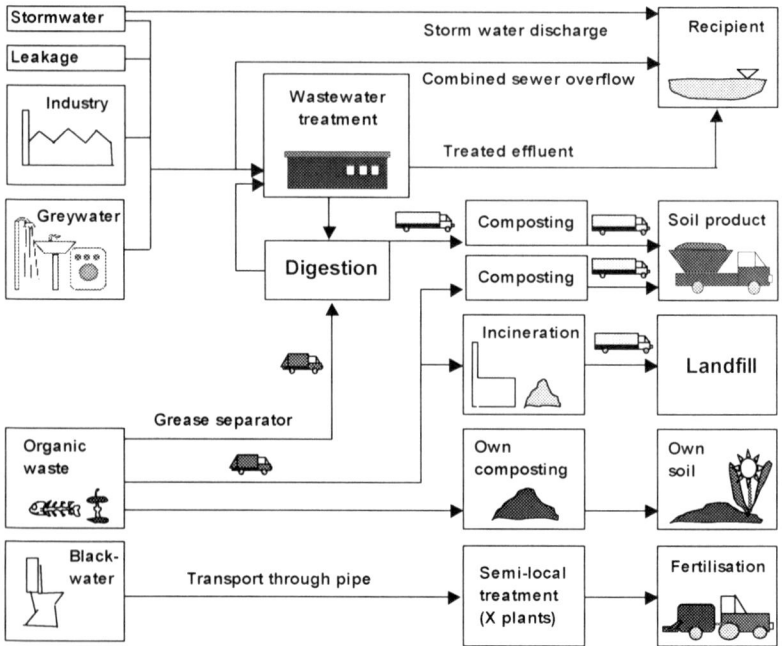

Figure 1-3. System alternative 3) Blackwater separation.

Comparing strategic options

In the phase of detailed evaluation, the system alternatives selected are simulated with the URWARE model, resulting in a quantitative output:

- The flows of nutrients, heavy metals, and some organic substances to the recipients, *i.e.* receiving waters, landfills, and arable land;
- The use of energy for treatment processes, and for transports; and
- The magnitude of global and local environmental impacts.

In separate studies, the stormwater system and the combined sewer system have been analysed with the SEWSYS model; these will be incorporated with the URWARE input data files. Further investigations are now being conducted on economy, health, receiving waters, technical function and household aspects. The results of the simulations and other investigations are to be structured in an easy-to-understand manner and used as a basis for a workshop, in which all available information can be discussed and integrated. This will be done by the project board, which includes the major stakeholders. The outcome of the workshop, subsequently disseminated to other stakeholders in the region, will be the foundation for decision-making.

Experience so far

At the point of writing, this ambitious project is about halfway towards its completion. The experience can be summarised as follows.

- The major stakeholders did not share the same views and goals at the beginning of the project, which delayed the initial phase. At the current stage, their views have converged, and the work in the project group runs smoothly. This underlines that the major stakeholders should try to find a common ground for the goals and ambitions of the project from start.
- The task of setting system boundaries requires consideration and agreement among the actors.
- The definition of future scenarios for 2050 is obviously complicated by the lack of any reliable data. Many assumptions and simplifications had to be agreed upon. Where the actors could not agree, a sensitivity analysis after the simulations should show whether the parameter in question is relevant for the outcome.
- The ambitions of the stakeholders have grown during the project, insofar as they wish to investigate more system alternatives. They also want to acquire real and detailed data for the simulations, instead of using the default values that are built into the models.
- The Urban Water approach and tools, which have so far proved useful and understandable, are being refined in the course of the project.

2

Assessment: environment and natural resources

2.1. Erik Kärrman, Andras Baky, Daniel Hellström,
*   Ulf Jeppsson and Håkan Jönsson*
2.2. Stefan Ahlman and Gilbert Svensson

2.1 URWARE: A MODEL FOR SUBSTANCE FLOW ANALYSIS

2.1.1 General description of the URban WAter REsearch model (URWARE)

As an important part of the Urban Water toolbox, it was decided to develop a systems analytical tool for assessment of environmental sustainability criteria of urban water systems, and to apply this tool to the model cities of the programme. Since the MATLAB-Simulink model ORganic WAste REsearch (ORWARE)

(Dalemo 1999) already existed which fulfilled some of the expectations of the requested model, it was decided to build the new model from the ORWARE structure. Due to the large number of necessary modifications the new tool was renamed the URban WAter REsearch model (URWARE), although the model development affects primarily the wastewater treatment part of the system.

URWARE can be described as a library of mathematical models, which are combined into an overall one of the specific physical system under investigation. URWARE includes both theoretical and empirical relations of substance flows. It handles average yearly data, *i.e.* the level of detail is low (see *e.g.* Hellström *et al.* 2000; Jeppsson and Hellström 2001). This allows sub-models to be simplified, variables to be aggregated and kinetic reactions to be disregarded. The present version of URWARE simulates 84 substances, providing non-aggregated outflows (emissions to air, soil and water) or aggregated environmental impacts of the system structure studied. A selection of the URWARE variables is given in Table 2-1.

The sub-models can be combined in any fashion, thereby generating various system structures that can be analysed and compared. The URWARE analysis is most often applied to relative comparisons between alternatives for the management of urban water flows for a given settlement or city. The models keep track of numerous substances related to wastewater, such as organic substances, nutrients and to some extent heavy metals. Within all models the release of environmentally hazardous substances to water, soil and air are calculated. The results can be used as one of many information pathways for strategic decision making related to urban water systems.

The sub-models describing the wastewater treatment processes include primary, secondary and final sedimentation units, thickener and dewatering units, sand filtration and biological reactors (anaerobic, anoxic and aerobic). These can be connected in any way and recycle flows added when required. The models are based on a detailed COD (chemical oxygen demand) description of the organic matter. In addition, total suspended solids (TSS), volatile solids (VS), total solids (TS) and biological oxygen demand (BOD) are calculated throughout the system. Five to seven nitrogen fractions (depending on the model) including nitrogen and N_2O gas production are maintained, as well as four sulphur (including sulphide), three phosphorus and two potassium fractions.

The physical mechanisms are modelled in a straightforward way, whereas biological reactions are described by an extended Activated Sludge model (ASM1) (Henze *et al.* 2000) with inspiration from both ASM2d and ASM3 models. The main extensions are related to phosphorus, sulphur and potassium. The fate of seven heavy metals is another modification. Although the complete URWARE model is primarily intended for steady-state analysis, the biological models describing the wastewater treatment plant are fully dynamic. The models also allow for chemical precipitation of phosphate, addition of external carbon sources, polymer addition, temperature dependency, calculation of energy consumption, etc. To enhance realistic behaviour of the models and assist the human user, every sub-model includes a significant number of on-line validation routines to ensure, for example, complete mass balances, a realistic choice of parameter values, consistent correlation between BOD, COD, VS, TSS and TS and reasonable nutrient contents of biomass. The details of the models and their behaviour are described in Jeppsson *et al.* (2005).

Table 2-1. A selection of the 84 variables in the URWARE input vector. The unit for each element is kg/year.

Notation	Parameter	Comment
H_2O	Water	
TS	Total solids	Fixed solids, ash = TS-VS
TSS	Total Suspended solids	Dissolved total solids
VS	Volatile solids	Organic material
COD_{tot}	COD, total	
$COD_{sol,bio}$	COD, soluble & biodegradable	
$COD_{sol,in}$	COD, soluble & inert	
$COD_{part,bio}$	COD, particulate & biodegradable	
$COD_{part,in}$	COD, particulate & inert	
BOD_7	BOD_7	Biochemical oxygen demand
N_{tot}	Total nitrogen	
$N_{NH3/NH4}$	Ammonia- and ammonium-nitrogen	
N_{NO3}	Nitrate-nitrogen	
$N_{sol,org}$	Nitrogen, soluble & organic	
$N_{part,org}$	Nitrogen, particulate& organic	
P_{tot}	Total phosphorus	
P_{PO4}	Phosphate-phosphorus	
P_{Part}	Particulate phosphorus	
S_{tot}	Total sulphur	
S_{SO4}	Sulphate-sulphur	
S_{S2-}	Sulphide-sulphur	
S_{part}	Particulate sulphurs	
K_{tot}	Total potassium	
K_{cell}	Potassium in cells	Dissociated K
Pb	Lead	
Cd	Cadmium	
Hg	Mercury	
Cu	Cupper	
Cr	Chromium	
Ni	Nickel	
Zn	Zinc	

The URWARE sub-models fit together with older ORWARE sub-models to form system structures for wastewater management. A compilation of sub-models from URWARE and ORWARE, useful for substance flow analysis of wastewater systems is given in Table 2-2.

Table 2-2. Main sub-models of URWARE and ORWARE with references.

Sub-model	Reference
Wastewater treatment plant	Jeppsson *et al.* (2005)
Sedimentation (incl. chemical precipitation), thickening, dewatering, sand filtration, activated sludge, anaerobic digestion)	
Urine separation systems	Jönsson *et al.* (2000)
Collection and transportation	Eriksson, O. *et al.* (2002)
Landfill	Björklund (1998)
Composting	Eriksson, O. *et al.* (2002)
Liquid composting	Kärrman *et al.* (1999)

In addition to the sub-models covered in Table 2-2 there are also simplified models for drinking water plants, pumping in sewer systems and irrigation by wastewater (Kärrman *et al.* 1999).

An example from a model city, Surahammar, is schematically given in Figures 2-1 to 2-3, showing three alternatives for the management of wastewater and organic solid household waste. They represent a conventional Swedish system, the existing system in Surahammar and a separating system primarily aimed at nutrient recovery.

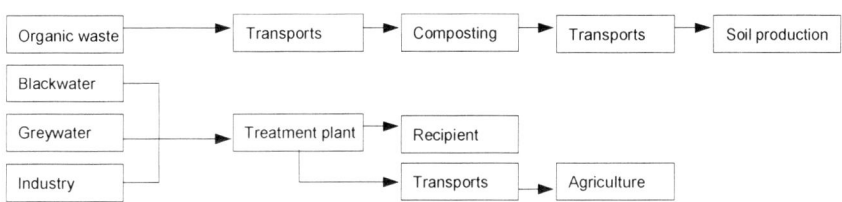

Figure 2-1. Conventional Swedish waste handling system (Alt. 1).

In Alternative 1, all wastewater is transferred through a sewer system to a wastewater treatment plant. The wastewater is treated mechanically, biologically and chemically before it is discharged to the receiving water. Sewage sludge is digested and transported to arable land to be used as a fertiliser. Solid organic waste is collected separately, transported to a composting plant and further used for soil production (Figure 2-1).

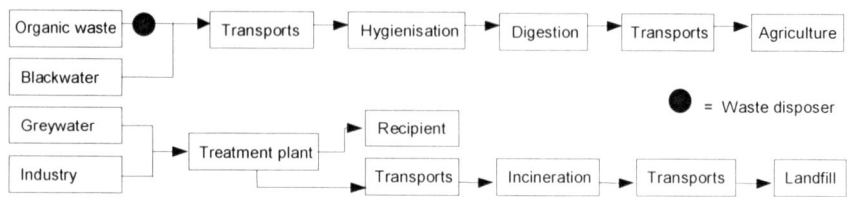

Figure 2-2. Existing waste handling system in Surahammar (Alt. 2).

In Alternative 2, there are three parallel systems for handling organic waste. A majority of the households have chosen food waste disposers from which food waste, together with the wastewater, is transferred by the sewer system to the treatment plant. The two alternative organic waste systems are collection for central composting (the same as Alt. 1) and home composting. The treatment plant is also operated as in Alt. 1. Sewage sludge is digested and transported to a soil-production facility (Figure 2-2).

Alternative 3 is formulated with the intention of generating a nutrient-rich product, suitable for recycling to agriculture, which originates only from urine, faeces and food waste (no mix with other waste or wastewater flows). Here, toilet waste from low-flush toilets and milled organic waste are collected in tanks and transported by trucks to a separate digestion plant from which the sludge is transported to arable land (without dewatering). Wastewater from baths, dishes and laundry (greywater) is transferred by sewers to the treatment plant for conventional treatment and the sludge produced is incinerated (Figure 2-3).

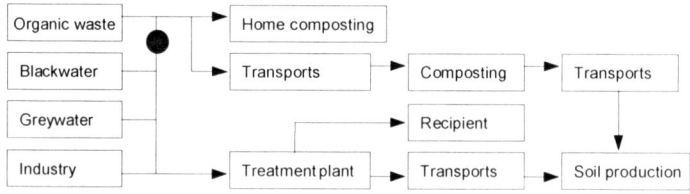

Figure 2-3. Blackwater system for nutrient recovery (Alt. 3).

2.1.2 Data needed

To analyse and compare new types of sewage systems, information about the composition of the different wastewater fractions is needed. In some systems the compostable household waste is mixed into the wastewater by use of a food waste disposer; hence, the composition of the solid compostable household waste is also needed. One of the stages in the development of URWARE was to compile well-validated default values on the mass and composition of urine, faeces, greywater and compostable household waste (Jönsson *et al.* 2005). The proposed default values were developed to fill the needs of the simulation model URWARE. However, the use of the proposed default values is not limited to URWARE simulations, since they can be used whenever default flows and composition of the described fractions are needed.

The URWARE input vector describes typical wastewater from households entering a wastewater treatment plant (WWTP). The influent wastewater is described as a mixture of urine, faeces and toilet paper, greywater and, in some systems, also biowaste from households. To obtain a description of incoming household wastewater, the flows and composition of urine, faeces, toilet paper, greywater and biowaste need to be known.

Table 2-3. The URWARE model household wastewater: a summation of urine, faeces and greywater, expressed as g pe^{-1}day^{-1}, mg/ pe^{-1} day^{-1} for the metals. Data quality: <u>well validated</u>, based on few references, *initial estimates* (Jönsson *et al.* 2005a).

Parameter	Urine	Faeces & toilet paper	Greywater total	Household ww total society
H_2O	1487	*110.6*		
TS	<u>20</u>	53.1	71.2	144.3
VS	7.4	46.4	41.6	95.40
COD_{tot}	*8.5*	*64.1*	62.4	*135.0*
BOD_7	5.0	*34.1*	33.8	*72.90*
N_{tot}	<u>11.0</u>	1.5	1.53	14.03
$N_{NH3/NH4}$	<u>10.3</u>	*0.3*	0.25	10.85
N_{NO3}	0	*0*	0.01	0.01
P_{tot}	<u>0.9</u>	0.5	0.68	2.08
P_{PO4}	*0.81*	*0.1*	*0.29*	*1.2*
S_{tot}	0.70	0.166	0.46	1.33
K_{tot}	<u>2.4</u>	0.9	0.79	4.09
Pb	0.012	*0.040*	1.3	1.35
Cd	<u>0.0005</u>	0.010	0.05	0.06
Hg	<u>0.00082</u>	0.009	0.005	0.01
Cu	0.10	1.10	10.3	11.50
Cr	0.010	0.13	1.3	1.44
Ni	0.011	0.19	1.6	1.80
Zn	0.3	*10.7*	13	*24.0*

Where local conditions differ significantly from average Swedish conditions, then modifications are certainly necessary, but for general base-line investigations of different treatment systems the given default vectors should be applicable. However, the vectors contain neither stormwater nor industrial wastewater; this means whenever these fractions enter the wastewater system, their compositions need to be found elsewhere. The default vectors defined are based on a thoroughly analysed literature study and a few additional measurements. Default flows and composition are given for urine, faeces, household greywater and compostable household waste. For these fractions the composition is described by 31 parameters, the flow of water, total and suspended solids, organic matter, BOD_7, total and fractioned COD, nitrogen, phosphorus, potassium and sulphur, as well as the flow of the heavy metals lead, cadmium, mercury, copper, chromium, nickel and zinc. A few main parameters for the proposed default vectors are given in Table 2-3.

2.1.2 Presentation of results

The URWARE model results are primarily substance flows, such as those to soil, water and air. Examples of these substance flows are: heavy metals to soil, COD and nutrients to water, and CO_2, SO_2 and NO_x to air. Figure 2-4 shows the simulated distribution of phosphorus in the three alternatives from Figures 2-1 to 2-3.

From Figure 2-4 it can be observed that, in alternative 1, the largest share of phosphorus (more than 80%) is expected to be used in soil, while almost 90% of the phosphorus would be used in soil production for ground constructions in alternative 2 (marked Left in material in Figure 2-4). In alternative 3 (blackwater system with nutrient recovery), around half of the phosphorus would end up in agriculture as recycled blackwater. Sludge from greywater treatment is assumed to be incinerated, and therefore almost 40% of the phosphorus ends up in ashes that are landfilled. All three alternatives have a share of around 10% that is discharged to receiving waters.

The URWARE model can also aggregate these substance flows into impacts such as eutrophication, global warming and acidification. These aggregations of impacts are made with methodologies from life cycle assessment (LCA). An example of the comparison of eutrophication is given in Figure 2-5.

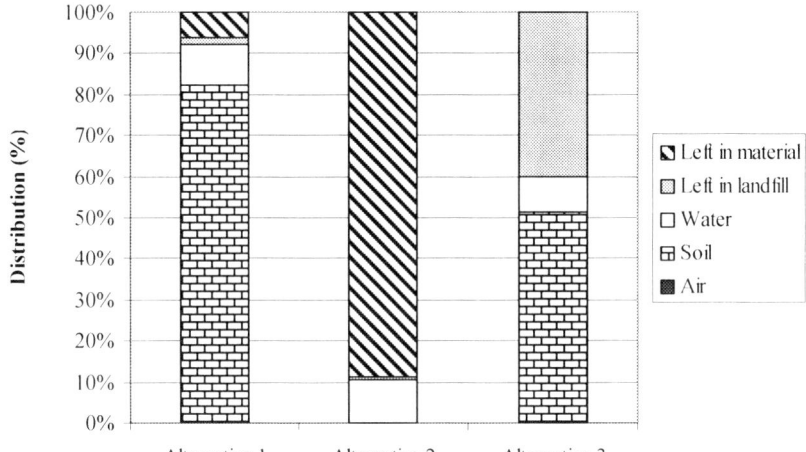

Figure 2-4. The percentage distribution of phosphorus in relation to the total inflow of phosphorus to the alternative systems in Surahammar.

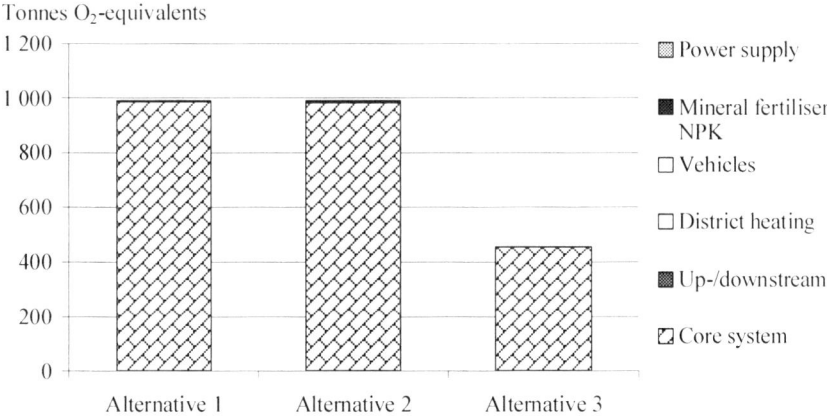

Figure 2-5. Example of a result from URWARE comparing eutrophication potential from the alternative systems in Surahammar.

The eutrophication potential expressed as oxygen consumption including primary consumption from nitrification and discharge of COD (Figure 2-5). The oxygen consumption also includes secondary effects from the discharge of phosphorus and nitrogen. The reason why alternative 3 has a smaller impact than the other two alternatives is mainly the lower amount of nitrogen discharged to the receiving water.

The URWARE model calculates and also sums up the energy use for all sub-models in the system studied. In addition, URWARE calculates the energy recovery from each of the alternatives, *e.g.* from biogas production and heat utilisation. Figure 2-6 shows the energy use and recovery in the three Surahammar alternatives.

MJ

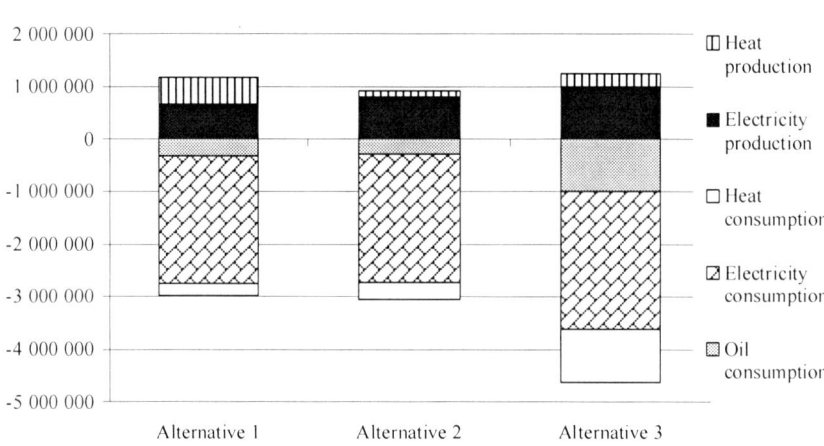

Figure 2-6. Energy use and energy recovery for the three alternatives in Surahammar.

One can see that it is possible to compare the alternatives in terms of total energy use, the total energy recovery as well as critical points in the system. In the studied case, the largest consumption of energy comes from process energy used for transport and treatment of wastewater fractions. In the current study, the biogas produced was assumed to be utilised in a gas engine for the electricity production.

2.1.4 Interpretation of the results

The URWARE provides the options to present results as substance flows, individual substances or as aggregated impacts such as eutrophication, global warming or acidification. However for decision-making or planning, even impact categories are not very easily interpreted which is why some expert guidance in setting priorities is needed. Three ways to assess the ecological dimension of sustainability were investigated by Hellström *et al.* (2004a):

- The guiding principles approach,
- Politically set environmental quality criteria, and
- Scientifically derived critical loads and carrying capacities.

Each of the approaches is used to measure the fulfilment of goals for urban water supply and wastewater management for six environmental criteria, namely water preservation, eutrophication, contribution to acidification, contribution to global warming, spreading of toxic compounds, and use of natural resources.

2.1.5 Model status and availability

The URWARE library of mathematical models is an expert tool in MATLAB/Simulink, and its application requires good knowledge of the model. Like the other tools that were developed within the Urban Water programme, URWARE is being developed further by applications in new and demanding projects, and by continued research. Under production is a simplified, Excel-based version that will be applicable also in sparsely populated areas. URWARE courses are arranged. Information on the latest version of the tool is available from www.urbanwater.org, or through contact with the authors.

2.2 SEWSYS: A DYNAMIC POLLUTION LOAD MODEL

The SEWSYS model is a source based substance flow model for simulation of transport and treatment processes in sewer systems, which was developed in the MATLAB/Simulink environment. SEWSYS consists of source modules for stormwater and sanitary wastewater, overflow structures and a conventional activated sludge wastewater treatment plant with phosphorus and nitrogen reduction. Stormwater can also be treated in a wet retention pond. The model can simulate both combined and separate sewer systems, as well as other source-separated systems with urine or blackwater diversion. The modelling framework in Figure 2-7 describes the stormwater part of SEWSYS.

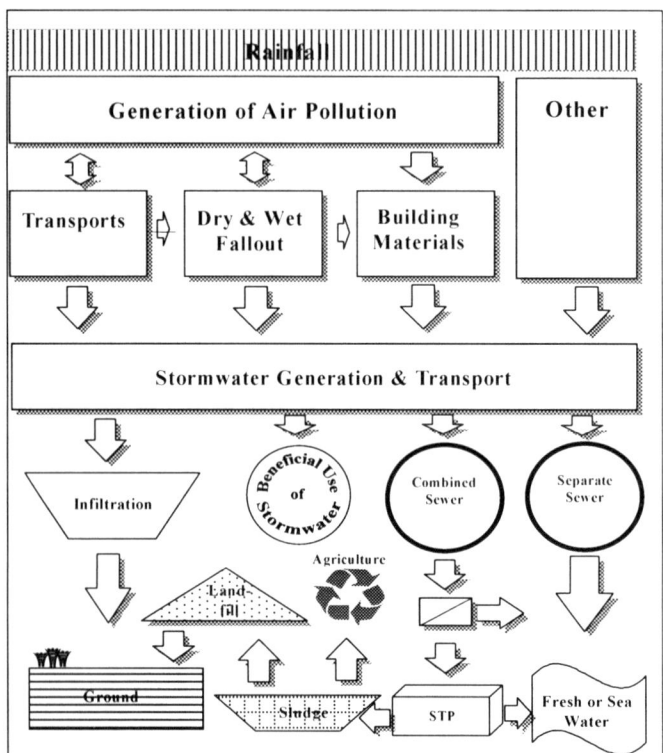

Figure 2-7. Conceptual modelling framework for the stormwater part of SEWSYS.

The modelling approach presented here also provides decision support for the implementation and evaluation of non-structural best management practices (BMPs). In this context non-structural BMPs consist of source control measures involving institutional, educational or pollution prevention practices.

2.2.1 Tool description

In SEWSYS the user can keep track of the pollutants, where they originate and where they finally end up. The Simulink model collects output data from the simulation in a set of matrices, which are stored in the MATLAB Workspace. These can be used for graphical display in MATLAB or be exported to other software such as Microsoft Excel. The elements of the matrices are mass flows gives as [g/s], except for water [m^3/s]. The model also has the capacity to calculate Event Mean Concentrations (EMCs) and the Site Mean Concentration (SMC). For sanitary wastewater, the pollution load can be derived from urine, faeces or greywater. Stormwater pollution can be identified from a range of sources such as atmospheric deposition, metal corrosion, brake wear, tyre wear and road wear.

The SEWSYS model consists of a main window, various input files consisting of background data on emission factors and Simulink models. The main window is the panel from which the user controls the model (Figure 2-8). This is where the user enters specific parameters for the catchment studied. When all of the parameters have been entered and the runoff routing process has been calibrated, the model is ready for simulations; the Simulink model then opens and runs with the specified parameters. The output can be viewed in plot windows or be post-processed in MATLAB or other software.

The main window also controls which type of sewer system, combined or separate, is to be used in the simulation. The choice of sewer system determines the Simulink file to be used in the simulation. It is also possible to specify a user-defined Simulink file. If the model is to run with a combined sewer overflow (CSO), the discharge threshold level in m^3/s must be specified.

There are two pre-defined Simulink models, one for combined systems and one for separate systems, since the fluxes take different paths in the two systems. In Figure 2-9, the top level for the separate system is shown. The Simulink model contains modules for pollutant sources, treatment modules and different recipients. There are four recipients: water, sewage sludge, landfill and air.

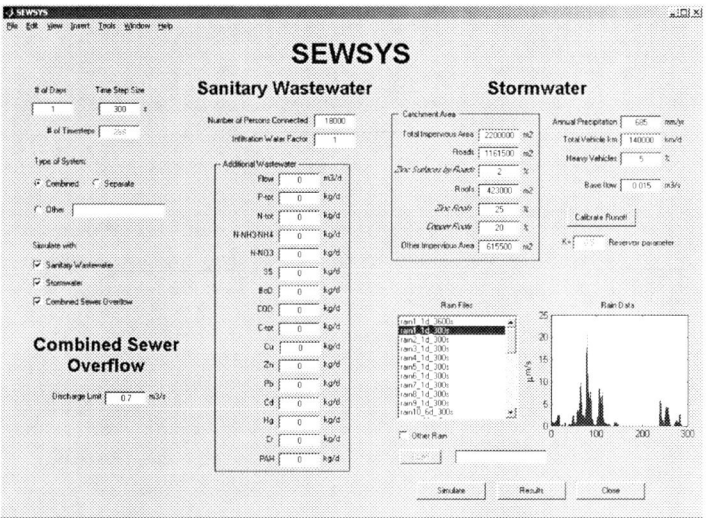

Figure 2-8. SEWSYS main window for data input and runoff calibration.

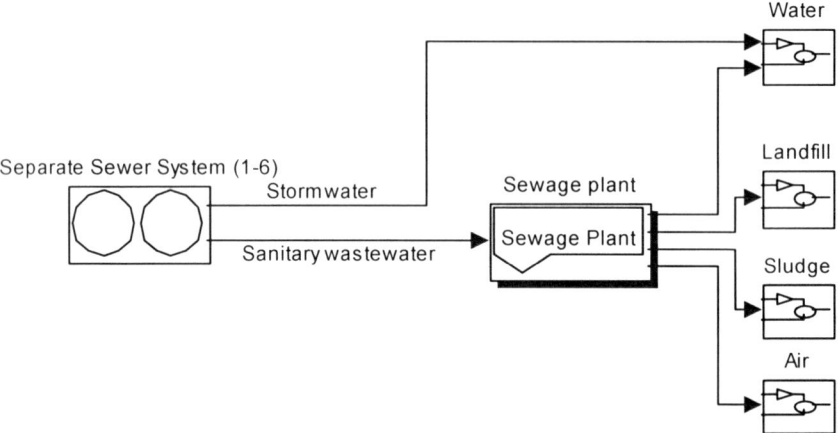

Figure 2-9. Top level for the separate system in SEWSYS-Simulink.

SEWSYS is a time discrete model with a fixed time step that can be set between 60 and 3600 seconds. For stormwater simulations the recommended time step is 5 to 15 minutes. The input that drives the stormwater simulation is a time-series of rainfall data that is up-loaded to the model. The rain data file has to be specified from the SEWSYS main window.

2.2.2 Data need

The parameters for sanitary wastewater consist of the number of people living in the catchment and the infiltration water factor. A factor of 1.0 means the same volume of infiltration water (*i.e.* drainage and erroneous connections) as sanitary wastewater. Background values for sanitary wastewater composition are given in per capita values in a separate input file. Additional wastewater, *e.g.* from industrial activities, may be added to the flow. The additional wastewater is integrated with the sanitary wastewater; consequently it has the same diurnal variation in composition and flow.

Parameters for stormwater include total impervious area, annual precipitation, traffic load and the percentage of heavy vehicles. The distribution between road surfaces, roof surfaces and other surfaces is also required for the impervious area. Surface areas of zinc (galvanised) and copper are also required. With the help of a GIS tool the characterisation of a catchment is made easier.

The model takes input from both the SEWSYS main window and the input files, known as m-files. The files *pollutants.m* and *constants.m* store input data for sanitary wastewater and stormwater emission factors, respectively. The emission factors for stormwater pollution include corrosion rates and emissions from different kinds of materials and traffic. Data for the pollutant sources in the stormwater and sanitary wastewater modules have been obtained from the literature as well as other existing Swedish models (Dalemo 1999; Stockholm Vatten 1999).

Table 2-4 shows the substance vector in SEWSYS. At present, these are the substances processed in SEWSYS. Other substances may be included, but that implies that background data of the new substance has to be available.

Table 2-4. The substance vector in SEWSYS.

No.	Substance	Unit	Comments
1	H_2O	[m^3/s]	Water
2	Tot-P	[g/s]	Total Phosphorus
3	Tot-N	[g/s]	Total Nitrogen
4	NH_3/NH_4^+-N	[g/s]	N in ammoniac and ammonium
5	NO_3 -N	[g/s]	Nitrogen in nitrate
6	N_2O -N	[g/s]	Nitrogen in nitrous oxide
7	SS	[g/s]	Suspended Solids
8	BOD_7	[g/s]	Biochemical Oxygen Demand (7 d)
9	COD	[g/s]	Chemical Oxygen Demand
10	Tot-C	[g/s]	Total Carbon
11	Phase Index	[-]	Part VS (Volatile Solids) of SS
12	Cu	[g/s]	Copper
13	Zn	[g/s]	Zinc
14	Pb	[g/s]	Lead
15	Cd	[g/s]	Cadmium
16	Hg	[g/s]	Mercury
17	Cr	[g/s]	Chromium
18	Pt	[g/s]	Platinum
19	Pd	[g/s]	Palladium
20	Rh	[g/s]	Rhodium
21	PAH	[g/s]	Polyaromatic Hydrocarbons

2.2.3 Using the tool

The first step in SEWSYS is to enter catchment specific data in the main window. Before the simulation can be started, calibration of the runoff routing module is needed. The calibration process is operated from the SEWSYS main window, the interface shown in Figure 2-10.

It is important that the runoff hydrograph be adequate since this determines the amount of water diverted from the combined sewer or emergency overflows. Calibration here means adjusting the parameter K for a given impervious area and a design rain with constant depth, so that the runoff is given a reasonable time of concentration, *i.e.* the time it takes for the whole area to contribute to the discharge at the endpoint. The parameter K is a function of Manning's number, the slope and runoff length and thus works as a damping coefficient; the runoff is damped and delayed.

The rain intensity used in the calibration should reflect the mean rain intensity of the current rainfall series to be used for the simulations. The reservoir parameter K should be seen as a pure calibration parameter. It represents the model's conceptual description of the routing of overland and pipe flow. The runoff hydrograph is the result from the calibration; if the curve is not satisfactory another run is made. Finally, the calibration process ends and sends the derived parameter K to the SEWSYS main window for subsequent model runs.

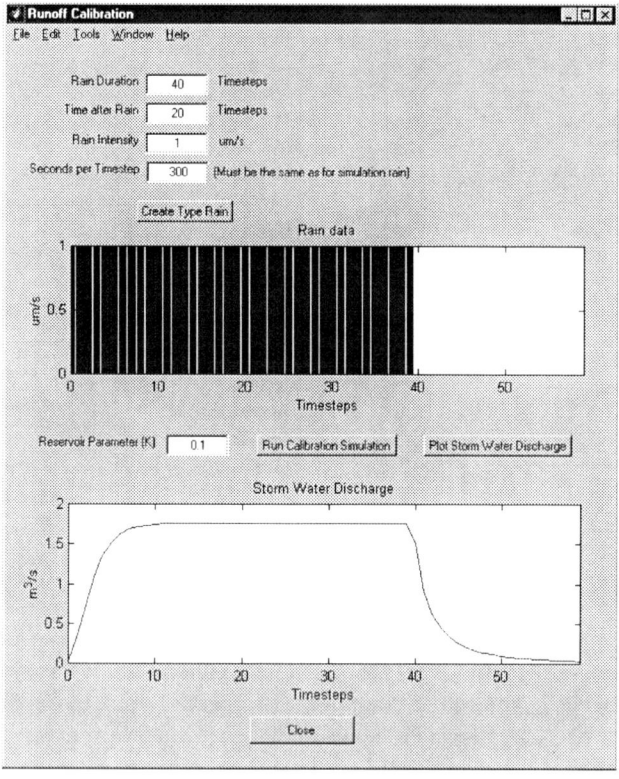

Figure 2-10. Calibration of the surface runoff from the SEWSYS main window.

Starting a simulation loads all input data and brings up the chosen Simulink model. The results can be accessed from the SEWSYS main window. Mass balances for the treatment plant, hydrographs, pollutographs, loadographs and sources for different pollutants may be displayed.

2.2.4 Presentation of results

The Simulink model stores data from each simulation in a set of matrices, which are placed in the MATLAB Workspace. The matrices normally have 21 columns, one for each substance, and the same number of rows as the number of time steps used in the simulation. The standard output matrices are compiled in Table 2-5.

Table 2-5. The output matrices in SEWSYS.

Matrix	Unit	Columns	Comments
s_urine	g/(capita*s)	21	Urine fraction
s_faeces	g/(capita*s)	21	Faeces fraction
s_bdt	g/(capita*s)	21	Bath- dishwash- and laundry fraction (greywater)
s_industry	g/s	21	Additional wastewater (from SEWSYS main window)
s_wetdep	$\mu g/m^2$	21	Wet deposition
s_road	μg	20	Pollutants from road areas
s_roof	μg	20	Pollutants from roof areas
s_other	μg	20	Pollutants from other areas
sewer	m^3/s or g/s	21	Total discharge and pollutants in sanitary wastewater
storm	m^3/s or g/s	21	Total discharge and pollutants in stormwater
inSewageplant	m^3/s or g/s	21	Influent to wastewater treatment plant
outCSO	m^3/s or g/s	21	Overflow from CSO
outRecipient	m^3/s or g/s	21	Emissions to water recipient
outWWTP_Recipient	m^3/s or g/s	21	Outflow from WWTP
outLandfill	m^3/s or g/s	21	Emissions to landfill
outSludge	m^3/s or g/s	21	Emissions to sludge
outAir	m^3/s or g/s	21	Emissions to air
storm_exceed	m^3/s or g/s	21	Bypass from overflow structure
storm2	m^3/s or g/s	21	Inflow to retention pond
non_settled	m^3/s or g/s	21	Outflow from retention pond
settled	m^3/s or g/s	21	Sediment fraction in pond

There are a few basic MATLAB commands, *i.e.* matrix operations that are useful when working with the output matrices. For example, if you want to know the total amount of copper in stormwater, generated during the simulation period, you need to work with the vector *storm* (Table 2-5). Since copper is number 12 in the SEWSYS substance vector (Table 2-4) the MATLAB command will look like this:

```
» sum(storm(:,12))*300
```

This command sums every row in column 12 and multiplies it by the time step size, in this case 300 seconds. Since column 12 has the unit [g/s], the answer is displayed in [g].

Pressing the Results button in the SEWSYS main window after a simulation (and possible loading of results) opens a Results Menu. From this menu the user can choose to display a selection of simulation data. Hydrographs, pollutographs, loadographs, total pollution loads, Event Mean Concentrations and source distributions can be displayed for both sanitary wastewater and stormwater.

2.2.5 Interpretation of the results

A range of potential uses of SEWSYS can be foreseen, mainly to provide decision support in studies of planning, scenarios and management by objectives. Another potential use is for education purposes, giving information and insights on the processes of generation and spreading of the harmful substances in the urban sewer system. A system analysis can provide useful information about the sustainability of different system structures, which then can be used as a basis for a decision-making process. An important part of the systems analysis is substance flow analysis. It gives valuable information about the pollutants in the system, where they originate, their transport and where they finally end up. When analysing aspects of sustainability for alternative wastewater systems each alternative system to be compared has to be modelled. The SEWSYS model has been used in a systems analysis of the wastewater system in a catchment called Vasastaden located in the city centre of Göteborg, Sweden (Section 9.3, Ahlman *et al.* 2004a; 2004b). In this study the model was used for calculating pollution loads and the sources of pollution. The results from the model were also used as input for an environmental impact assessment. The Simulink models in Figures 2-11 and 2-12 show the present sewer system in Vasastaden, Göteborg, and an alternative sewer system, respectively.

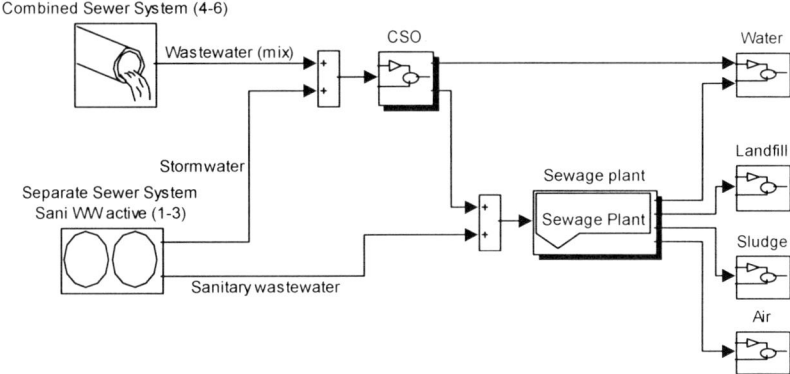

Figure 2-11. Matlab/Simulink model for the present sewer system.

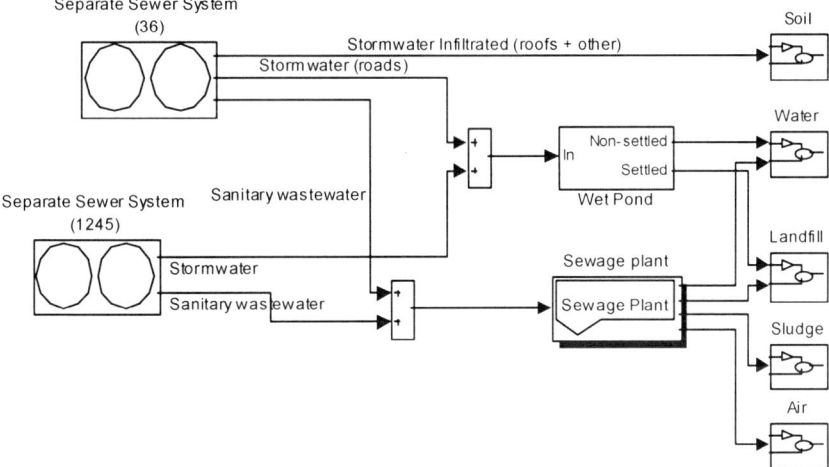

Figure 2-12. Matlab/Simulink model for a separate sewer system with source control measures and stormwater treatment.

The results from the substance flow analysis in Vasastaden are shown in Figure 2-13. It is clear that the largest part of phosphorus and nitrogen can be found in urine and faeces; phosphorus and nitrogen in the urine fraction is 50% and 80%, respectively. Greywater is a large source for heavy metals and BOD whereas stormwater contributes mainly to heavy metals and PAH.

SEWSYS also has the ability to show the distribution of pollution sources for stormwater to a higher degree as shown in Figure 2-14. In Vasastaden, corrosion of roofing material and brake wear are the major sources for copper in stormwater. Zinc has more contributing sources; corrosion of roofing material and tyre wear are the major sources for zinc pollution. For lead, cadmium, phosphorus and nitrogen atmospheric deposition (wet and dry) is the dominating source. PAH in stormwater comes mainly from tyre and road wear.

In studies where scenarios are compared, bar graphs such as Figure 2-15 are useful. This graph comes from a study where SEWSYS was used to simulate different strategies for management of polluted stormwater from an urban highway in Göteborg, Sweden (Svensson et al. 2001). The first column from the left shows the amount of copper if no measures are taken. The other columns represent the amount of copper for four abatement strategies. Similar graphs can be made for all pollutant sources, and by comparing them the most efficient abatement strategy can be found.

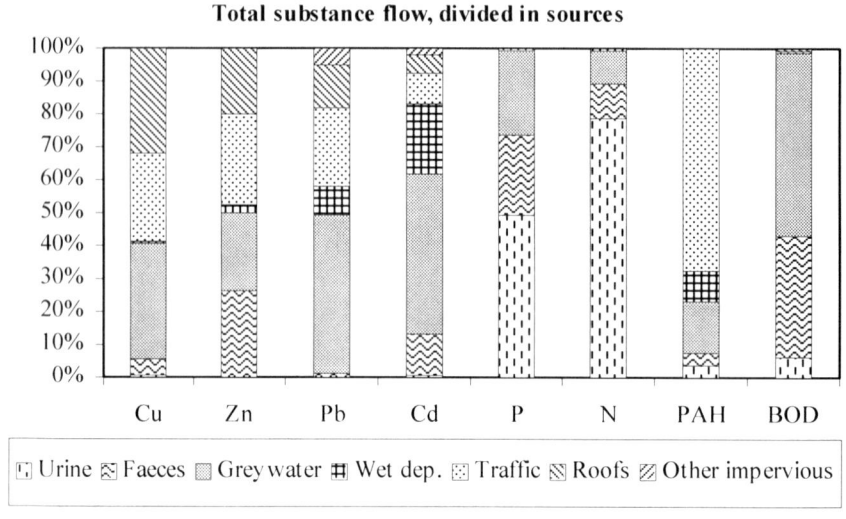

Figure 2-13. Total substance flows in Vasastaden divided in sources for sanitary wastewater (urine, faeces and greywater) and stormwater (roads, roofs and other impervious areas).

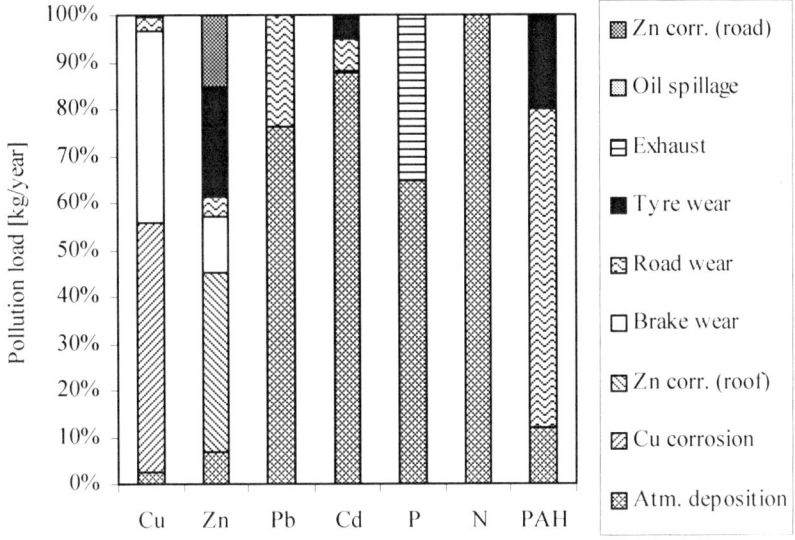

Figure 2-14. Distribution of pollution sources for stormwater in Vasastaden.

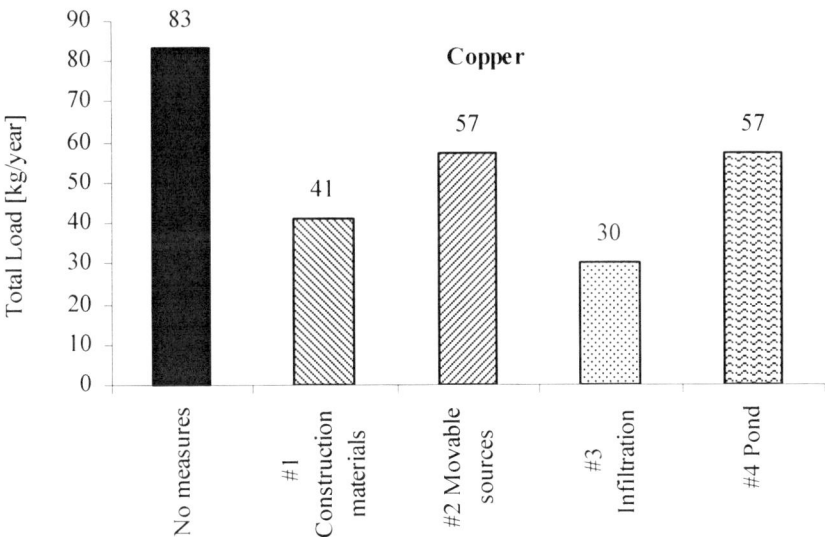

Figure 2-15. Copper mass flows to the receiving water in the four scenarios.

It is concluded here that the substance flow model SEWSYS is suitable for quantifying the pollutant loads from diffuse sources in an urban catchment. The model also gives a deeper understanding of the processes of release and transport of the substances included. Alternative scenarios of abatement strategies can readily be simulated, the results evaluated, and the pollution sources identified.

2.2.6 Model status and availability

The SEWSYS model is designed to run in a MATLAB environment. The minimum requirements for running SEWSYS are MATLAB version 6.5 and Simulink version 5.0, running on a computer with at least 256 MB RAM. For information on the model's availability the reader is directed to the SEWSYS section on the website of Urban Water (www.urbanwater.org). In future versions of SEWSYS, hazardous organic substances will be included.

2.2.7 Further reading

The underpinning-modelling concept in SEWSYS is the result of two Master's degree projects at Chalmers University of Technology (Ahlman 2000) and Uppsala University School of Engineering (Engvall 1999). Further refinement of the model was done in Stefan Ahlman's Ph.D. project (Ahlman and Svensson 2002). The most recent and detailed description of the tool is given by Ahlman and Svensson (2005).

SEWSYS has previously been used in several studies to simulate substance flows in urban sewer systems (Section 9.3, Svensson *et al.* 2001; Ahlman *et al.* 2004a; 2004b; 2005a). In these studies SEWSYS was able to categorise the sources and sinks for wastewater and stormwater pollution in different types of sewer systems. In addition, the model has been used to evaluate scenarios for source control measures.

3

Assessment: hygiene and health

*3.1. Nicholas Ashbolt, Susan Petterson,
   Therese Westrell, Jakob Ottoson,
   Caroline Schönning and Thor Axel Stenström*
*3.2. Anna Ledin, Eva Eriksson, Peter Mikkelsen and
   Anders Baun*
3.3. Helena Almqvist

3.1 MICROBIAL RISK ASSESSMENT (MRA) TOOL TO AID IN THE SELECTION OF SUSTAINABLE URBAN WATER SYSTEMS

Traditionally, the hygiene impact of urban water systems has been assessed by the occurrence of thermotolerant coliforms and E. coli indicator bacteria at

© IWA Publishing 2006. *Strategic Planning of Sustainable Urban Water Management* edited by Per-Arne Malmqvist, Gerald Heinicke, Erik Kärrman, Thor Axel Stenström and Gilbert Svensson. ISBN: 1843391058. Published by IWA Publishing, London, UK.

points of exposure or treatment. However, reliance on faecal indicator bacteria is simply inadequate to cover the range of treatment-resistant pathogens potentially released from modern urban water systems (Ashbolt *et al.* 2001). International guidelines for waters (Bonn Charter 2004; WHO 2003; WHO 2004) strongly advocate an assessment of the performance of barriers to pathogens from sources, through the water supply system and to recipients. Nonetheless, suitable tools are not yet generally available to support agencies in assessing pathogen risks, along with environmental, social and technical aspects in the context of a systems analysis for sustainability assessment.

The development of a system's lifetime approach to applying quantitative microbial risk assessment (QMRA) is described here (the MRA tool). In the future, different urban water system options will be compared on the basis of infections per month (to account for outbreak potential), the variation or uncertainty in this infection risk (system robustness), and identify critical pathogen(s) and pathway(s) of infection, to aid risk management.

The overall criterion for sustainability, considered in the MRA tool, from a human hygiene perspective, is that the risk of infection, directly or indirectly, from environmental sources should never exceed a minimal background level. The acceptable background level may, however, differ for various regions of the world and over time, and local data on known infection rates can be amended by users of the tool. The first criterion assessed is the ability of a system structure to provide an acceptably low infection level, currently considered to be <1 per 10 000 people per annum, as proposed by the US-EPA and Dutch regulation for microbial risks from drinking water (Anonymous 2001; Regli *et al.* 1991).

The MRA tool provides hygiene information appropriate for integration with the other projects in the Urban Water Programme, and it is based on what has been previously published by the consortium (Malmqvist *et al.* 2000). In taking a systems approach the MRA tool attempts to account for system failures and intrinsic variability of unit operations over their intended life-times; in theory it can be applied to a broad range of urban water systems and it continues to be developed by the authors.

3.1.1   MRA tool description

The MRA tool is based on information collected from various small communities within Sweden (Gebers, Vibyåsen and Hammarby Sjöstad in Stockholm, and Vasastaden in Göteborg) and data collected from Stockholm's largest wastewater treatment works. New system configurations will be added in future versions, and local parameter values may have to be added if disease rates are significantly different from those in Scandinavia. System alternatives are compared by the MRA tool, based on the average infections likely to result per system per annum and, to assess robustness (and outbreak potential), the number of extreme infection events likely over the lifetime of the system. The formal procedure of Quantitative Microbial Risk Assessment (QMRA, Haas *et al.* 1999) is used, but in a novel way to compare whole systems over their lifetimes. The approach is therefore termed life-cycle pathogen risk assessment.The four major elements of QMRA are:

- Hazard identification – general identification of pathogens that can cause infection or disease (interpreted by a limited range of reference pathogens);
- Exposure assessment – when, in what dose and how many people are likely to be exposed to a pathogen?
- Dose-response assessment (or hazard characterisation) – establishment of a relationship between the dose of a microbial agent and the rate of infection; and
- Risk characterisation – the result of the analysis presented as a numerical value, most often as infections per 10 000 individuals per annum, or for recreational waters as infections per 1 000 swims.

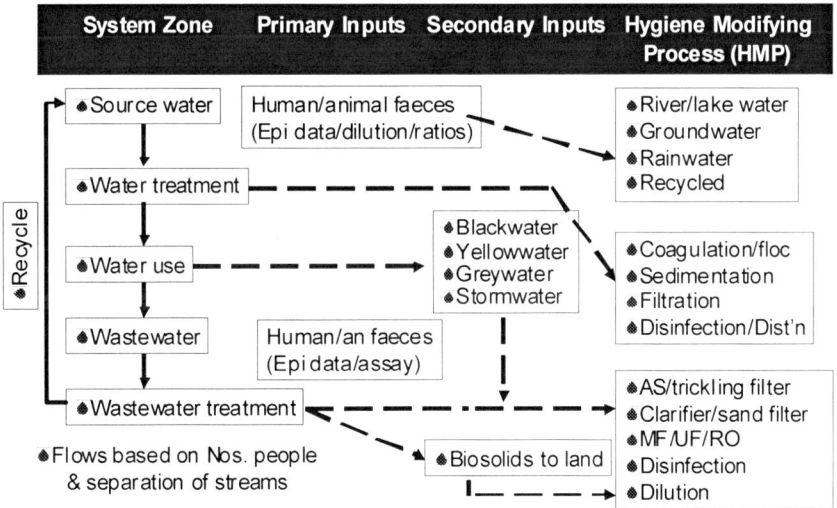

Figure 3-1. General schematic of the MRA tool for human health assessment. Epi = epidemiologic; Dist'n = distribution.

Each system is subdivided into its Hygiene Modifying Processes (HMPs). The MRA tool is designed to allow user selection of system components, which activates pre-defined, specific datasets for pathogen system performance (removal and inactivation) and reliability, coded in the software package Analytica® (Version 3, Lumina Decision Systems Inc., Los Gatos, California). Each unit operation in the system structure has pre-described variability and pathogen performance data, *e.g.* for sand filter, chlorination or distribution (Figure 3-1). Users then confirm likely points of exposure (*e.g.* in home, recreation water site or drinking water). Dose-response models for the reference pathogens were used as described by Haas *et al.* (1999) using relevant exponential and beta-Poisson models. Selected viral, bacterial, parasitic protozoan and helminth (worm) pathogens relevant for the transmission routes serve as index organisms. A specific strength of the approach has been the use of pathogen index data to estimate exposure(s) to the reference pathogens. Probability distributions are utilised rather than point estimates (means), to better account for the large variation in numbers of pathogens encountered. Randomly selected reference pathogen numbers are selected from their distributions (Monte Carlo simulations) and run through each treatment barrier (HMP) to produce a final distribution of pathogens at points of exposure.

Ideally, pathogen changes within each HMP, due to treatment efficiency, variability and failure rates (including potential growth for bacterial pathogens), are expressed as probability distributions in order to account for spatial and temporal variability. However, since available data on performance and failure levels are limited, guestimates are used or left open for the user to add. These could then provide further quantification of the impact of failure or the effect of using a more reliable HMP. Previous qualitative and quantitative sensitivity analyses were used to identify the parameters and pathways that have the largest influence on the risks, thus limiting the parameters for which detailed analysis is required. The Analytica® software provides influence diagrams which are hierarchical, providing the user with an intuitive means to see increasing detail, as illustrated in Figures 3-4, 3-5 and 3-6. Estimated infections for the life-time of a system structure are compared by running Monte Carlo simulations of pathogen numbers through each system thousands of times, so as to capture acute effects which may lead to outbreaks. System robustness is estimated based on the range in variability and uncertainty over the system life and normalised to annual infections. The MRA tool also provides valuable information for risk management. Rankings of infection risk potential by exposure sites enables the identification of control points for pathogen management, within the Hazard Analysis Critical Control Points (HACCP) risk management approach promoted by WHO and other water guideline setting agencies (Bonn Charter 2004; WHO 2003; WHO 2004).

As a tool to aid in selecting treatment options, alternative systems can be compared based on relative levels of infection risk, while acknowledging the uncertainties in each. Hence, existing and hypothetical systems may be compared and, when integrated with other assessments (environmental, economic, social and technological), provide management the breadth of information necessary to evaluate system sustainability. During a multi-criteria decision making process, the MRA tool output can further inform the various stakeholders involved.

3.1.2 Data need and pathogen data sources

Risk assessments can be undertaken at various levels of precision and sophistication. An initial coarse screening (designed to eliminate low risks and focus on key risks) is normally undertaken with conservative assumptions and generic data. This first approach is often referred to as a screening-level or Tier 1 risk assessment. The approach adopted for the MRA tools is a Tier 2 QMRA, which uses probabilistic Monte Carlo simulation with basic summary probability functions (*e.g.* triangular, normal, uniform) reflecting local water quality and operational measurements.

Epidemiologic and literature data along with direct measurements in the Swedish systems were utilised to estimate viral, bacterial, and parasitic protozoan pathogen ranges in source materials (faeces, urine and sewage). Pathogen removal by the key system units (described above) was modelled by reference pathogen numbers (rotavirus, adenovirus, *Campylobacter*, *Cryptosporidium* and *Giardia*). Reference pathogen numbers in source faeces, urine and sewage were described as probability density functions (PDFs in Table 3-1) and their dilution, removal and inactivation described as the pathogens passed through each system treatment structure (HMP). Overall, helminth risks were considered low, but were included from estimates in sewage (with *Ascaris lumbricoides* as the reference pathogen).

A potential limitation in the Tier 2 QMRA arises from the lack of quantitative data on various pathogen groups in source materials and the inability, with present laboratory methods, to quantitatively detect some pathogens at the level of concern in treated waters. Hence, these concerns have been addressed by two approaches. One is to estimate numbers from the probability of individuals being infected with key organisms from each group of pathogens. The required data is derived from epidemiologic studies and estimates of pathogen excretion rates. The second approach utilises pathogen surrogates, such as particle or spore removal across barriers, or faecal biomarkers to assay faecal cross-contamination when actual pathogen numbers are likely to be very low, highly variable, or both. All of these approaches are illustrated in our papers (Fane and Ashbolt 2000; Höglund *et al.* 2002a; 2002b; Westrell *et al.* 2003; Westrell *et al.* 2004) and incorporated into the MRA tool database in Analytica®.

From concern that our epidemiologic approach (Table 3-1) may overestimate pathogens in faeces and sewage, results were compared with sewage datasets from

Australia (taking into account minor differences in the prevalence of reported disease). Overall there was good agreement (within an order of magnitude) for the reference pathogens, but higher variability was exhibited by the epidemiologic approach than by the assayed one (from data collected over one year, see Figure 3-2. Virus numbers in sewage, however, were estimated to be some three orders of magnitude higher than numbers cultured (infective) from sewage, which may reflect reality (virus recovery was not reported, but could be 10 to 40%). Furthermore, non-culturable viruses such as Norovirus, have been reported by non-culture (i.e. PCR methods) at levels similar to those reported in Table 3-1 (Lodder et al. 1999). Therefore both virus estimations were used in the MRA tool to provide a sense of the uncertainty in the overall estimation of viral infection risks.

Table 3-1. Epidemiologic data used to estimate pathogen probability density functions (PDFs) in sewage (modified from Höglund 2002b).

Parameter	Enteric virus (Rotavirus)	Bacterium (C. jejuni)	Protozoan (C. parvum)	Protozoan (G. lamblia)
Yearly incidence of infection (%)	0.95	15.6	0.31	0.84
Disease rate if infected (%)	75[a]	23	39	50
Log_{10} Excretion time (days)	N[b] (1.0,0.30)	N (1.18,0.325)	N (1.48,0.173)	N (1.18,0.325)
Log_{10} Excretion density (no/g faeces)	N (10, 1)	N (8, 1)	N (7, 1)	N (7, 1)
Estimated number in sewage[c]				
Mean (l^{-1})	1.53×10^7	3.38×10^6	1.27×10^4	2.13×10^4
50th percentile (l^{-1})	8.97×10^5	1.87×10^5	8.97×10^2	1.24×10^3
95th percentile l^{-1})	4.55×10^7	1.03×10^7	4.13×10^4	6.29×10^4

[a] Only for children and the aged (assumed to be 15% of population)
[b] Normal distribution defined by parameters (mean, standard deviation)
[c] Assuming 150g faeces/(person*day) and 135 litres /(person*day) water released to the sewer.

In our previous quantitative microbial risk assessments of urban water systems (Höglund et al. 2002b; Ottoson and Stenström 2003a; Westrell 2004; Westrell et al. 2003; 2004), a wide range of sites with potential exposure to pathogens was investigated. The MRA tool makes use of this knowledge and limits potential exposure sites to only those considered more likely and important, however, the exposure sites studied may be chosen by the user. Examples include sites A to C in Figure 3-3, but the point of biosolids application, given the previous pasteurisation and thermophilic digestion of the sludge, was considered far lower and not further assessed.

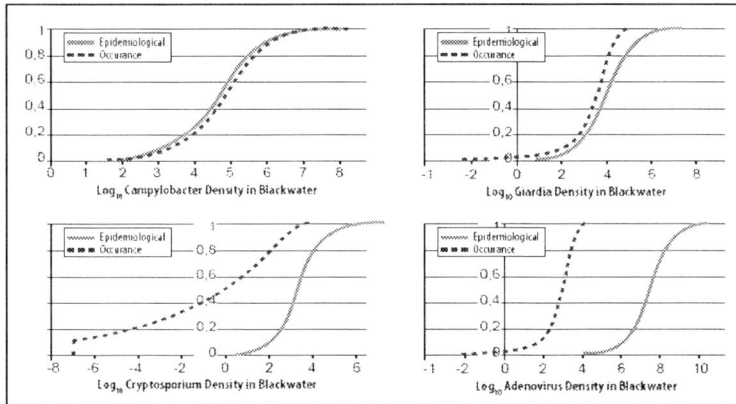

Figure 3-2. Example of output from MRA tool illustrating spread in the pathogen PDFs when estimating numbers (per litre) in sewage from epidemiologic datasets (solid line, see Table 3-1) or occurrence datasets (dashed line).

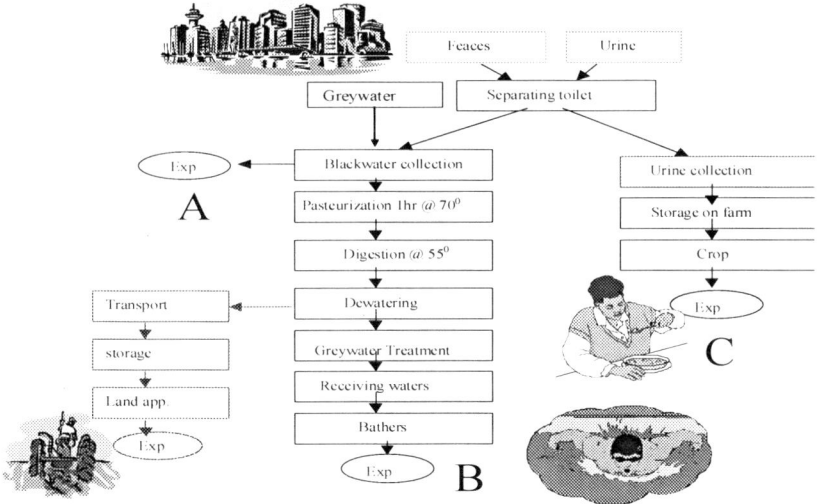

Figure 3-3. Source-separation case. Possible exposure (EXP) points of concern identified: A) workers fixing plumbing & sewage overflow; B) recreational water users in receiving water; and C) consumers of crops grown with urine fertiliser.

3.1.3 Using the tool

In the following, the application of the tool is exemplified in a comparison of two wastewater systems in an Urban Water model city (see also Section 9.1).

System Structures

For the centralised structure, wastewater was conveyed by sewer to one large activated sludge plant (with screening, grit chamber, pre-aeration, pre-sedimentation with coagulant, activated-sludge with nitrification and denitrification, and coagulation-sand filtration). There was no disinfection of the effluent. The sludge was processed by pasteurisation, anaerobic thermophilic digestion, and the KREPRO (Kemwater, Helsingborg, Sweden) nutrient recovery process prior to soil application of biosolids. A flow diagram of the Source-separation system is provided in Figure 3-3, with the main difference being that urine was separated by urine-diversion toilets, stored (for disinfection), and used as a fertilizer for crop production.

MRA Tool screens

The initial page of the MRA tool allows for the selection of system components in each system-structure, typically two or more to be compared in what are called influence diagrams (Figure 3-4). An example of user inputs for combined (centralised) and source-separated sewage systems is illustrated in Figure 3-5. Exposures are then calculated for each pathogen group (Figure 3-6).

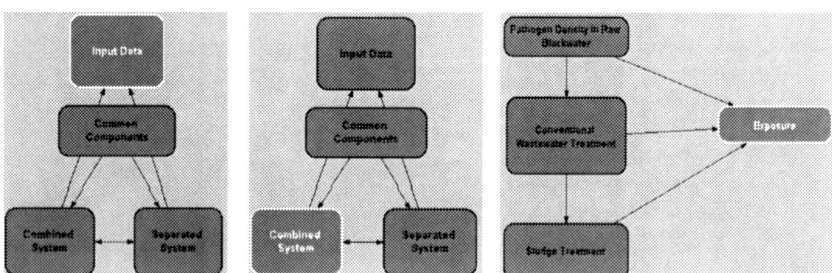

Figure 3-4. Initial screen of MRA tool to compare combined and separated system alternatives (left, described in Section 9.1) with Input Data (Figure 3-5) and combined system (centre) opened revealing further influence diagrams within (right).

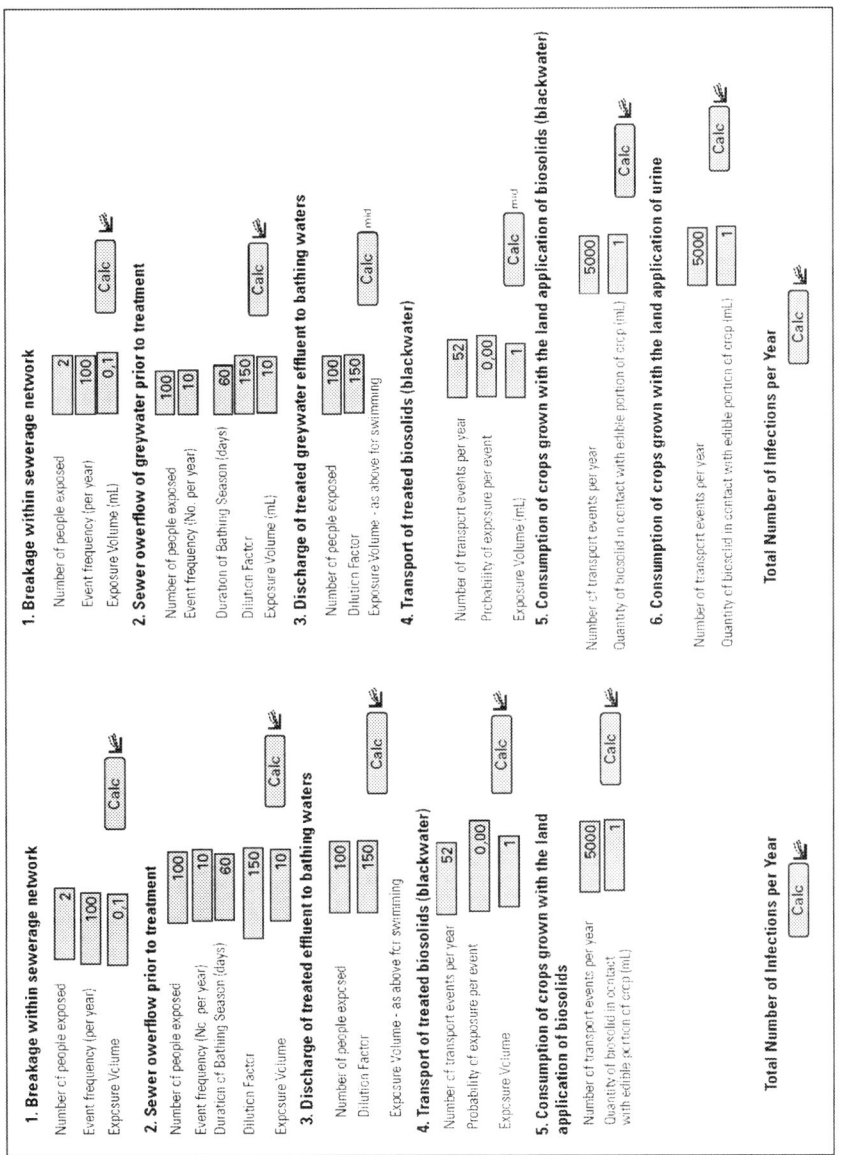

Figure 3-5. Example of an influence diagram in the Input Data (influence diagram illustrated top left, 4) used to assess the combined System (left) and separated (right). Each input has default values, which can be user-changed and the effect is shown within the 'Calc' boxes.

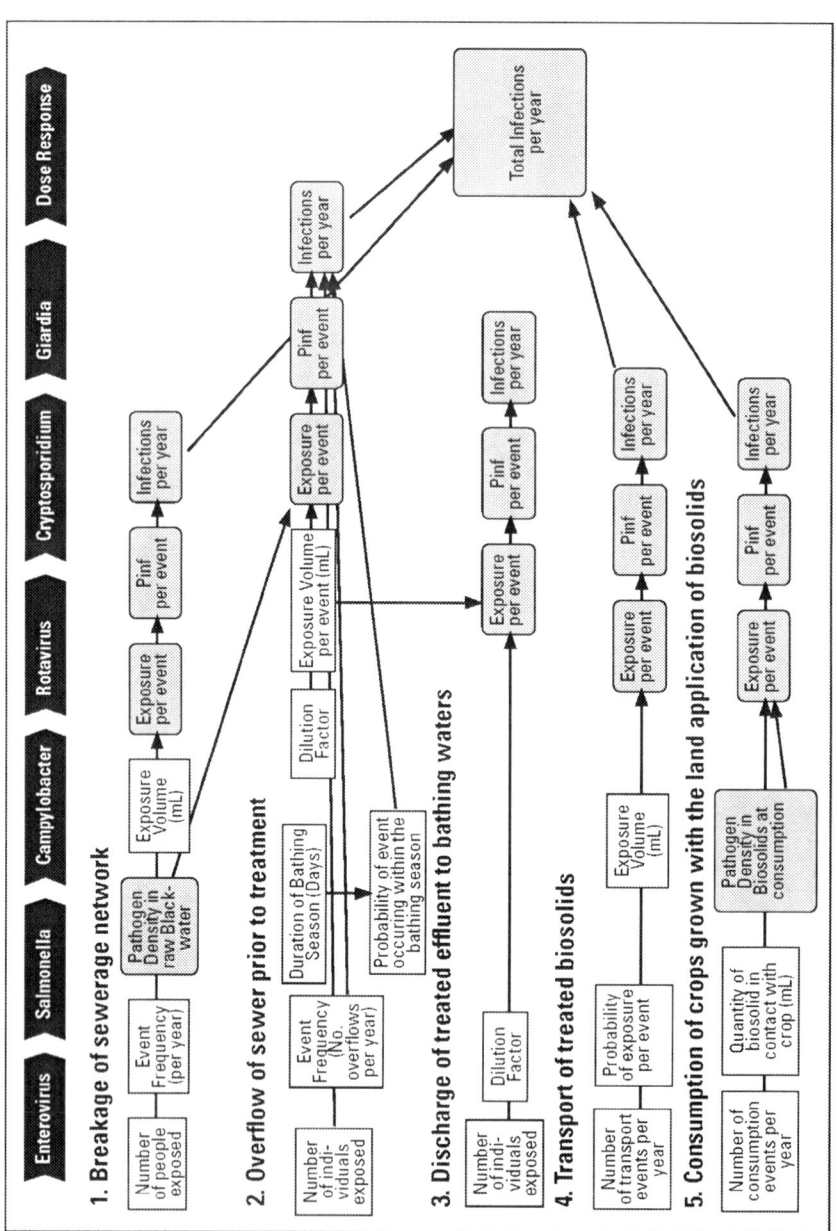

Figure 3-6. Example influence diagram of various exposures accounted for in the MRA tool (source separated system structure) P_{inf} = probability of infection.

3.1.4 Presentation of the results

Results can be viewed in two ways, the first of which is interactively within Analytica® to allow the user to flick between the results of each system structure selected. Alternatively, results can be exported to an Excel spreadsheet for further data manipulation. Examples of the results from the analysis of the model city investigated are presented in Figure 3-7.

After performing the calculations, the model output needs to be put into context. To interpret the results, a valuation of disease rates has been proposed (Table 3-2) to which model outputs may be compared.

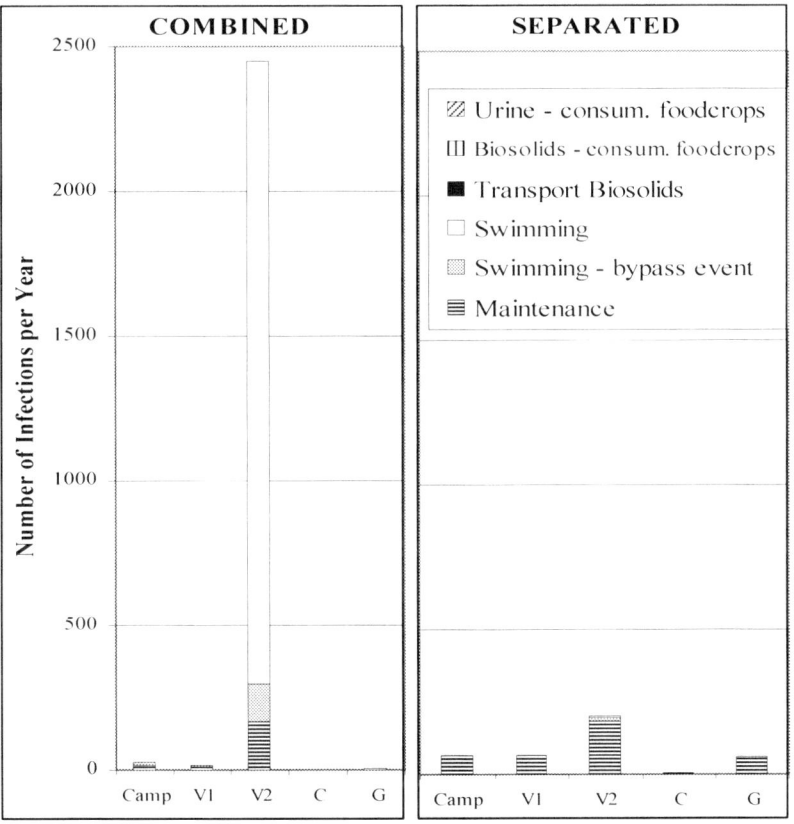

Figure 3-7. Estimated annual infections for Combined (left) and Separated (right) system structures for reference pathogens (camp: *Campylobacter*, V1 & V2: different assumption for virus excretion density (V1=Log_{10} (Triangular[1,8,10]) and V2= Log_{10} (Normal[10,1]) in Table 3-1), showing sensitivity, C: *Cryptosporidium*, G: *Giardia*).

Table 3-2. Risk rankings based on estimated disease rates (from Westrell et al. 2004)

Item	Definition
Catastrophic	Major increase in diarrhoeal disease >25% or >5% increase in more severe disease or large community outbreak (100 cases) or death
Major	Increase in more severe diseases (0.1-5%) or large increase in diarrhoeal disease (5-<25%)
Moderate	Increase in diarrhoeal disease (1-<5%)
Minor	Slight increase in diarrhoeal diseases (0.1-<1%)
Insignificant	No increase in disease incidence (<0.1%)

The following paragraphs may exemplify the type of interpretation that the tool provides.

The pathogen group of highest concern varied by point of exposure, but based on overall risk, enteric viruses generally dictated the outcome in both the centralised and source-separated alternatives, not only in the current example but also in other system structures studied to date (Höglund et al. 2002b; Ottoson and Stenström 2003a; Westrell et al. 2003; 2004). The highest likelihood of infection stemmed from obvious risks, such as exposure to raw sewage during mains breaks or leaks in the centralised alternative, or unblocking urine-diversion pipes in the source-separated one; yet few individuals would be exposed during such scenarios. Overall community risks were ranked minor for both systems, but the source-separate system yielded the lower risks, due to less potential impact from recreation swimming. Given the potential significance of viral infections, management of aerosols becomes an important aspect to workers near raw wastewater streams.

Another important method of controlling viruses is sufficient storage of urine before crop application (from one month for cereals to six months for vegetables) at temperatures of about 20°C (Höglund et al. 2002a). Disinfection of wastewater discharged to receiving waters would also have improved the performance of the centralised system, which would then leave maintenance exposures as the main residual issue for both systems.

3.1.5 Model status and availability

At the time of writing, the MRA tool was in a beta testing stage, with only a limited number of selectable system structures in place. It is envisaged that more structures will be added. In summary, the MRA tool is designed to assist both expert system designers and generalist stakeholders who are involved in multi-criteria decision aided selection of system options. The MRA will enable them to become better informed of the hygiene risks associated with urban water systems. The model is considered a living tool that will develop with each application of new system information and use. One of the advantages of the software package is that there is a freely available version for running constructed models, such as the MRA tool. It is envisaged that a free version of the MRA tool built in Analytica® software will be made available in the medium term. Currently, beta testing of specific systems is available through discussion with the authors. A detailed description of the MRA tool has been published recently (Ashbolt *et al.* 2005)

3.2 CHIAT: CHEMICAL HAZARD IDENTIFICATION AND ASSESSMENT TOOL FOR SELECTION OF PRIORITY POLLUTANTS

For the evaluation of alternatives in wastewater handling, it is necessary to assess potential hazards and problems related to chemicals. The types of hazards and problems that should be taken into account are related to the strategy selected for handling. Hazards due to exposure to chemicals of humans, livestock, aquatic and terrestrial organisms, crops and plants have to be considered, as well as technical and aesthetic problems. Hazards that need to be taken into account are both acute and long-term effects on living organisms, *e.g.* toxicity, bioaccumulation, carcinogenicity, mutagenicity, reproduction hazards and endocrine disrupting effects, as well as promoting allergic reactions in humans. Precipitation of salts and minerals and corrosion of some types of installations and tubing are examples of relevant technical problems. Bad odour, frothing and colouring are examples of aesthetic problems to be considered.

A screening procedure, CHIAT, Chemical Hazard Identification and Assessment tool, for selection of the most critical pollutants involved in strategies for handling of storm- and wastewater is introduced here. CHIAT is a procedure for identifying and assessing chemical hazards and problems with the aim to identify priority pollutants. The tool comprises 5 steps, some of which are computer models, some are procedures, and one is a roadmap for obtaining a constructive discussion that leads to a decision. The work presented in this section is, due to space limitations, restricted to one category of pollutants the xenobiotic organic compounds (XOCs), to two types of polluted waters, stormwater and greywater, and to one strategy for handling, discharge of storm- and greywater to a surface water recipient after settling of suspended solids. For more details, see Ledin *et al.* (2005).

3.2.1 Methodology

Risk assessment of chemicals is composed of four elements: hazard and problem identification, hazard and problem assessment, risk characterisation and risk management, according to the technical guidance document (TGD) for risk assessment of chemicals in the EU (European Commission 2003). In general, hazard identification serves to map the inherent properties of chemicals by collecting and comparing relevant data; *e.g.* physical state, volatility and mobility as well as potential for degradation, bioaccumulation and toxicity. Hazard assessment is divided between exposure assessment and effect assessment. Comprehensive model systems have been developed to assess the distribution of contaminants in the environment (soil, water, air) and in tissue (animals, humans). The next step is risk characterisation, where the potential negative effects are evaluated and, if possible, the probability of such effects occurring is estimated. Finally, risk management involves a range of possible interventions, *i.e.* monitoring and control of emissions to reduce risk environments (see Mikkelsen *et al.* 2001).

The methodology developed in this work is inspired by the TGD (European Commission 2003) and consists of five steps: 1) Source characterisation, 2) Recipient, receptor and criteria identification, 3) Hazard and problem identification, 4) Hazard assessment, and 5) Expert judgement, (Figure 3-8).

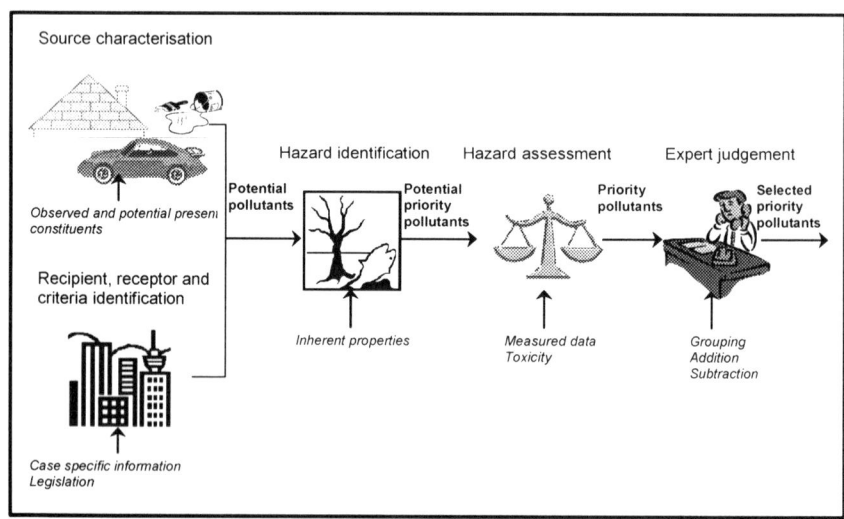

Figure 3-8. An approach to selecting priority pollutants based on chemical hazard identification and assessment.

Source characterisation

Initially, information regarding the potential pollutants should be collected. This should be done by two somewhat different approaches:

(1) Searching in the open international literature for observations and measurements of XOCs in the different types of wastewater. This is generating a list of observed constituents (for more details, see Eriksson, E. *et al.* 2002, 2003, 2006, and Ledin *et al.* 2005).

(2) Searching in the literature for XOCs potentially present in storm- and wastewater due to: a) The use of chemical products such as household chemicals (*e.g.* shampoo, toothpaste, cleaning detergents and washing powders), pharmaceuticals, de-icing agents, and pesticides in gardens. b) Releases from materials such as building and road materials, brakes and tires on vehicles; c) Atmospheric deposition; and d) Food and drinking water (Figure 3-9). The aim is to identify those pollutants that have not yet been included in monitoring programs. For more details, see Eriksson, E. *et al.* (2005), Ledin *et al.* (2004; 2005).

The lists generated in Ledin *et al.* (2005) can be used for other projects. However, it should be kept in mind that some compounds may have to be added. There could be XOCs that are used in special cases, such as de-icing agents and pesticides in stormwater, and medicines in urine and faeces. The output from this step is a list of all observed and potentially present pollutants: the potential pollutants (Figure 3-9).

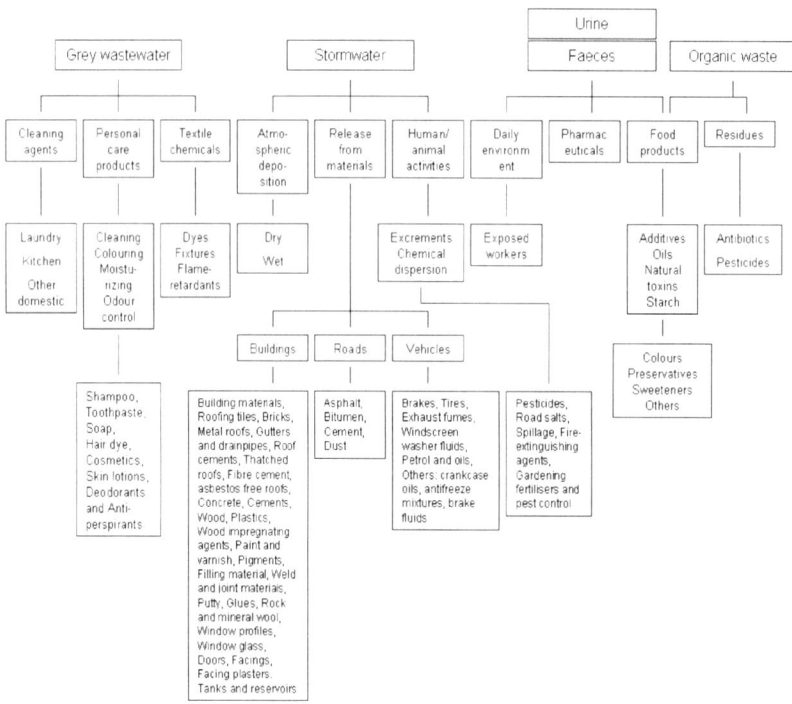

Figure 3-9. Examples of sources that are potential contributors to pollutants in wastewater fractions.

Recipient, receptor and criteria identification

In this step the strategies for handling of the selected type of waste- and stormwater are evaluated with respect to potential human health hazards, technical and aesthetic problems as well as environmental hazards that could occur due to the presence of the chemical pollutants.

Table 3-3. Examples of recipients, receptors and related criteria for hazard assessment.

Recipient	Receptor and hazard or problem	Criteria for hazard assessment
Air	Exposure of humans: allergic reactions, cancer, mutagenic effects, reproduction and toxicity	Human dose-response analysis
Ground water	Exposure of humans: allergic reactions, cancer, mutagenic effects, reproduction and toxicity	Human dose-response analysis Drinking water quality standards
	Exposure of livestock	Animal dose-response analysis
	Effects on the aquatic ecosystem when groundwater enters surface water	PEC/PNEC^a >1 in water Environmental quality standards Emission limit values
Installations	Clogging	K_s/IAP >1^b
	Colouring	K_s/IAP >1
	Corrosion	
	Frothing	
	Microbial growth	N & P discharge quality standards
	Taste and odour (T&O)	PEC or T&O threshold >1
	Precipitation	K_s/IAP >1
Soil	Effects on the soil ecosystem	PEC/PNEC >1 in soil
	Uptake by plants	Soil quality standards
	Exposure of livestock	Animal dose-response analysis
	Exposure of humans: from eating the crops: allergic reactions, cancer, mutagenic effects, reproduction and toxicity	Human dose-response analysis
Surface water	Effects on the aquatic ecosystem	PEC/PNEC >1 in water Environmental quality standards Emission limit values
	Exposure of humans: allergic reactions, cancer, mutagenic effects, reproduction and toxicity	Human dose-response analysis Drinking water quality standards
	Exposure of livestock	Animal dose-response analysis

^{a} PEC = predicted environmental concentration (PEC); PNEC = predicted no effect concentration.
^{b} Ks = solubility constant for the mineral or salt; IAP = Ion Activity Product for the relevant ions.

The usage, treatment or discharge scenarios are investigated in order to identify the potential exposure routes and what or who is exposed (the receptors). Legislation should be reviewed, for each scenario, to elucidate the quality and emission standards to be used in steps 4 and 5. Examples of recipients of relevance are water, soil and technical installations. Humans as well as aquatic and other terrestrial organisms are examples of receptors (Table 3-3). Exposure routes for humans are oral intake, inhalation of aerosols and skin contact. Both continuous and pulse emissions need to be taken into account when evaluating discharges to the aquatic environment and applications to soil. The selection of criteria for hazard assessment is discussed below. The output from this step is a list of recipients, receptors and criteria for hazard assessment.

Hazard and problem identification

All constituents identified as potential pollutants in the first step of the procedure (Figure 3-8) are evaluated in the third step, the hazard and problem identification. The criteria for evaluation are based on environmental fate (sorption, volatilisation, persistence, and bioaccumulation), short-term aquatic toxicity and long-term effects on living organisms (carcinogenicity, mutagenicity, reproduction damage and endocrine disrupting effects as well as promoting allergic reactions in humans).

The hazard and problem identification is made according to a ranking methodology, RICH (Ranking and Identification of Chemical Hazards), which was developed and tested for XOCs (for more details see Baun *et al.* 2005). The methodology consists of a decision tree in which hazardous and problematic compounds are identified. To visualize the sorting of XOCs, the decision tree can be described as a funnel fitted with several filters. The filters were set according to specified criteria based on sorption, volatility, persistence, potential for bioaccumulation and aquatic toxicity. There are also on/off filters for technical or aesthetic problems, as well as a long-term chronic effects filter for cancer, mutagenic and reproduction hazards, endocrine disruption effects and allergenic effects. The output is a classification of the compounds in three categories (white, grey and black) depending on their priority as possible pollutants. White compounds are considered non-priority pollutants, which means that these compounds are excluded from the fourth step, the hazard assessment. Grey compounds are passed on to the next filter. These compounds may or may not be priority pollutants depending on the outcome of the following filtration. Black compounds are considered priority pollutants.

The first filter is designed to separate compounds into water phase compounds and solid phase compounds. In this case the underlying assumption is that the water is transported in open systems facilitating good contact between air and water, *i.e.* highly volatile compounds are identified as white compounds.

It should be noted that no compound would be designated as black as a result of this first filtration. The purpose of this filter is to label the XOCs according to their preference for the water or the solid phase (*e.g.* suspended solids, sediment and soil). This information will be used for the further evaluation in step 4, the hazard assessment. Information required, in the present study, can mainly be collected from databases and handbooks presenting the inherent properties of the XOCs (for more information and references, see Baun *et al.* 2005). It is expected that decision trees modified for metals and other inorganic constituents will be developed in the future.

The inherent properties of the potential pollutants needed to carry out this step have to be compiled from the literature (databases and handbooks). The compounds that are identified as hazardous or problematic in this step are listed as potential priority pollutants.

Hazard assessment

The hazard assessment is, in general, a comparison between the effect and the exposure levels of a specific pollutant, made to evaluate whether there is an actual hazard under the present circumstances. The criteria selected for a hazard assessment are related to the receptor (Table 3-3). The exposure can be represented by predicted environmental concentrations (PEC), for which the values can be based on measured data or model simulations. Evaluation of estimated concentrations at which unacceptable effects are not likely to occur can be made by estimating predicted no effect concentrations (PNEC). The values can found in the literature (databases and handbooks), or they can be estimated using toxicity data collected for step 3, applying recommendations by the European Commission (2003). Comparison of the PEC and PNEC values are made to determine whether the compound should be considered hazardous for organisms in the environment. The pollutants for which PEC/PNEC ratios are above one (1) are classified as priority pollutants. A corresponding evaluation with respect to humans can for example be performed by using tolerable daily intake (TDI) values, to retrieve effect-values.

Expert judgement

Finally, the expert judgement is made. The expert is not necessarily a single person (*e.g.* an environmental chemist) but may be a group of decision-makers with different backgrounds. Due to financial limitations in a given project, the expert judgement may aim to reduce the number of compounds on which further action can be taken. Compounds may be removed from the list based on usage patterns in the catchment, or grouped based on similarity in structure and fate. An indicator compound may be chosen to represent the whole group.

Compounds that are banned can also be removed unless further prevalence in the environment is suspected.

Legislation concerning limits, *e.g.* drinking water standards, as well as environmental quality standards and emission standards for watercourses, lakes or the sea, reviewed in step 2 (Figure 3-8 and Table 3-3), should be used to identify compounds that may need to be added to the list. Compounds may also be added if they are priority pollutants present on national or international lists or special case compounds. The output from this step is a list containing those chemical compounds and other parameters that constitute a hazard after evaluation by the expert: the selected priority pollutants.

3.2.2 Two examples: greywater and stormwater

Two types of wastewater were chosen to illustrate how the screening procedure works, greywater and stormwater. The focus is on xenobiotic organic compounds (XOCs), since the third step in the procedure, the hazard and problem identification has up to now been developed for XOCs. One strategy for handling is considered: the discharge of greywater and stormwater to a surface water recipient, after settling of solid matter.

Source characterisation

In studies by Eriksson, E. *et al.* (2003) and Palmquist (2004b) a total of 246 XOCs have been identified in greywater from bathrooms (showers and hand basins). Several fragrances such as citronellol, coumarin and hexylcinnamicaldehyde were identified, as well as some preservatives *e.g.* parabenes and triclosan. The measurements also showed that biologically active chemicals (pharmaceuticals) were present, as well as unexpected compounds not directly derived from household chemicals, such as flame-retardants. The presence of detergents, softeners and preservatives was also confirmed. The XOCs that can potentially be present in greywater were listed, based on information available on chemical products, consumption statistics and chemical databases (for more details, see Eriksson, E. *et al.* 2002, 2003, and Ledin *et al.* 2005). The total list includes 867 compounds and compounds groups. Relatively few XOCs (36) fall in both categories, *i.e.* have actually been found in greywater and were identified as potentially present based on knowledge of the presence of the compound in specific household chemicals.

The literature survey of observations and measurements of pollutants in stormwater showed that a large number of constituents have been identified and quantified, in total, 366 XOCs were found (Eriksson, E. *et al.* 2006). The number of the potentially present XOCs was 411. It should be mentioned that the searching carried out within the project was limited (for more details see

Ledin *et al.* 2005). There was a relatively limited number of compounds (121) that belonged to both groups, *i.e.* compounds that have been identified in stormwater and pointed out as potentially present. This observation indicates that although a large number of organic compounds have been observed in stormwater, there could be at least as many other compounds present for which no one has yet analysed.

Recipient, receptor and criteria identification

In the scenario for handling of greywater and stormwater, discharge to a surface water recipient after settling of solid matter, the recipient was a hypothetical water body, with an assumed dilution factor of 100. In this scenario aquatic organisms are to be protected against toxic effects, and they are accordingly the receptors. Furthermore, a relevant criterion for the hazard evaluation is the ratio of PEC to PNEC, where PEC is the concentration of the pollutant in the lake and PNEC is the concentration below which no negative effects can be expected on aquatic organisms.

Hazard and problem identification

It was found that at least 1077 XOCs could potentially be present in greywater. So far, 205 of them have been evaluated according to step 3, the hazard and problem identification procedure described above (Ledin *et al.* 2005). Of these 205 XOCs, 24 were classified as hazardous in the water phase and 46 in the solid phase.

Furthermore, at least 656 XOCs could potentially have an impact on the stormwater quality. So far, 189 XOCs have been evaluated (Ledin *et al.* 2005); 65 of these 189 XOCs were classified as hazardous in the water phase, whereas 85 were classified as hazardous in the solid phase.

Hazard assessment

Measured concentrations found in the review by Eriksson, E. *et al.* (2006) were used in the present study to represent PEC-values in stormwater, while measured concentrations in untreated greywater were used for the greywater example (Eriksson, E. *et al.* 2003). It was possible to find PEC values for all 24 XOCs in greywater, while the corresponding value for stormwater was 43. PNEC values could be found for all 43 stormwater XOCs, while only 17 greywater XOCs could be evaluated due to lack of PNEC values.

The PEC to PNEC ratio exceeded the value of 1 for three XOCs, in greywater discharged to surface water, after settling of solid matter and associated XOCs. These were a softener (Dibutylphthalate), a pesticide (Malathion) and a preservative (Pentachlorophenol).

The PEC to PNEC ratio exceeded the value of 1 for 17 XOCs, for stormwater discharged to the hypothetical lake. Among them were 14 pesticides: Acrolein, Atrazine, Chlordane, Dichlorprop, Dichlorvos, Dieldrin, Diuron, Endosulfan, Malathion, Metazachlor, Methyl-Parathion, Propiconazol, Simazine and Terbutyazine. Two PAHs were also pointed out: Benzo[a]antracene and Naphthalene.

More information on concentration ranges in the different types of wastewater, efficiency of treatment methods, fate in receiving waters and soils, and data for the effect analysis is needed in order to refine this step.

Expert judgement

Finally, the most important XOCs are chosen for the list of selected priority pollutants.

In the present example the following points were considered for **greywater**.

Legislation

No XOCs need to be added to the list of selected priority pollutants for greywater due to legislation.

Other lists of priority pollutants

It should be decided whether some compounds need to be considered if they should be added due to their presence on EU's Water Framework Directive priority list (European Commission 2004). These are Nonyl phenol ethoxylates (NPEO) and Octyl phenol ethoxylates (OPEO, see Table 3-4). They are added to the list because, besides being on EU's list of priority pollutants, they are also potentially present in greywater and have been identified as potentially hazardous in water in step 3, the hazard and problem identification. For more details, see Ledin *et al.* (2005).

Exclusion due identical sources or similar environmental properties

There are no arguments for excluding any of the compounds identified in step 4, hazard assessment, since the three compounds represent three separate types of compounds.

Accordingly, 1077 potentially present XOCs were evaluated and the result for greywater is a list of 5 selected priority pollutants (Table 3-4).

The following points were considered for **stormwater**.

Legislation
No XOC needs to be added to the list of selected priority pollutants for greywater due to legislation.

Other lists with priority pollutants
It should be considered whether some compounds need to be added due to there presence on EU's Water Framework Directive priority list (European Commission 2004). The only one of interest was Hexachlororcyclohexane (HCH), since it has been identified as potentially present in step 1, source characterisation, it was classified as a potential priority pollutant in water in step 3, hazard and problem identification, while it has not been evaluated in step 4, hazard assessment, due to lack of data.

Excluding due to e.g. the same sources or similar environmental properties
For stormwater, the XOCs present on the list that was generated in step 4, hazard assessment, can be grouped according to the following (Ledin *et al.* 2005).

(a) PAHs: benzo(a)antracene and Naphthalene should both be included on the list of selected priority pollutants, since they have different inherent properties and can represent the whole group of PAHs.

(b) Chlorotriazine group (herbicides): Atrazine, Simazine and Terbutylazine. Terbutylazine was chosen since both Atrazine and Simazine are banned.

(c) Alkene group (herbicides): Acrolein was selected since it is the only representative of this group.

(d) Chlorinated aliphates or aromates (pesticides): Chlordane, Dieldrine, Endosulfane, Metoxychlor. Metoxychlor was selected, since the others are banned.

(e) Phenoxy acids (herbicides): Dichlorprop, was selected since it is the only representative of this group.

(f) Organo phosphates (insecticides): Dichlorvos, Malathion, Methyl-parathion. Diklorvos was selected since Methyl-parathion is banned and the data used for calculation of PEC for Malathion are based on just one study.

(g) Organic nitrogen (herbicides): Diuron was selected since it is the only representative of this group.

(h) Chloro acetanilides (pesticides): Metazachlor was selected since it is the only representative of this group.

(i)   Azoles (fungicides): Propiconazol was selected since it is the only representative of this group.

Accordingly, 656 potentially present XOCs were evaluated, and the result for stormwater is a list of 11 selected priority pollutants (Table 3-4).

Table 3-4. List of selected priority pollutants

Greywater	Stormwater
Dibuthylphthalate	Acrolein
Malathion	Benzo [a] antracene
Nonyl Phenol Ethoxylates	Dichlorprop
Octyl Phenol Ethoxylates	Dichlorvos
Penta Chloro Phenol	Diuron
	Hexachlorocyclohexane
	Metazachlor
	Methoxychlor
	Naphthalene
	Propiconazol
	Terbuthylazin

The large numbers of XOCs (656) that potentially are present in stormwater were reduced to 11 selected priority pollutants. However, some of the reduction was due to lack of data, *e.g.* it was not possible to evaluate all 656 compounds either in step 3, due to lack of data regarding inherent properties, or in step 4, due to lack of data regarding environmental concentrations (PEC). Corresponding numbers for greywater were 1077 potentially present pollutants reduced to 5 selected priority pollutants. Data were lacking for a full hazard and problem identification, and also for predicted no effect concentrations (PNEC).

3.2.3 Tool status and availability

The CHIAT tool developed was found to be very promising. The major advantages are:

- It can be used generally for identifying priority pollutants to evaluate alternative strategies for handling of storm- and wastewater.
- It can be used for selection of priority pollutants to be included in monitoring programmes.
- This procedure for selecting pollutants is transparent and adaptive to the specific scenario in focus.

A more detailed description of how to use the CHIAT procedure is available through contact with the author. The CHIAT tool is meant to be developed further in cooperation with its users. Three examples of areas needing further development are given.

- The database generated in step 1, source characterisation, has to include more XOCs. It is the authors' strong belief that the number of compounds identified as potentially present would greatly increase, if the search for information continued. It is therefore important, that the users of CHIAT add compounds relevant for their specific case to step 1, source characterisation.
- For the third step in CHIAT, hazard and problem identification, RICH is available as a β-version programmed in Excel, and it will be available in the future as a separate tool. Also here, more data on inherent properties is needed.
- The fourth step, hazard assessment, needs to be developed further to include a function that helps the user to estimate PEC and PNEC values for different recipients and objects of exposure.

Information on the latest version of the tool is available on www.urbanwater.org or through contact with the authors.

3.3 BARRIERS TO PREVENT HAZARDOUS SUBSTANCES IN WASTEWATER SYSTEMS

The effluents of wastewater treatment plants have been shown to be significant pathways for hazardous substances to enter the aquatic environment (Daughton and Ternes 1999; Heberer 2002). A complete mineralisation of xenobiotic compounds in treatment systems is rare, with the term biotransformation more accurately describing the potential changes of such compounds (Byrns 2001). Some compounds are biotransformed into harmless products and become degraded. Other compounds form metabolites that may be more or less toxic than the parent, while still others may prove to be generally recalcitrant and persist within the treatment plant (Byrns 2001; Heberer 2002). Hazardous substances that resist degradation in WWTPs may remain in the effluent, and thus are emitted to the receiving water, or enter the sludge. In areas where surface water partly originating from the effluent wastewater is used as a raw water source for drinking water production, concentrations of these substances may build up over time and pose a risk to human health (van der Voet *et al.* 2004). Hazardous substances that escape water and wastewater treatment continue to raise increasing concerns, especially regarding their potential effects in water ecosystems and on human fertility (SIWI 2004).

The transport of hazardous substances with wastewater is a critical issue not only for the receiving waters but also for sludge management. The content of hazardous elements and organic substances in wastewater and sludge reflects a society where many different chemicals are used in large amounts and for a variety of purposes. Their occurrence in wastewater is dynamic; it depends on the lifestyle and behaviour of all system users (Kroiss 2004). This feature makes the reuse of wastewater sludge on arable land a controversial issue in many countries where it is a widely discussed and debated topic (Kroiss 2004; Rulkens 2004).

In brief, there is disagreement between two main parties, one of which promotes the reuse of sewage sludge, thereby simultaneously resolving a waste problem and managing the recycling of biosolids and nutrients to arable land; if not recycled, these substances must be replaced by mineral fertilisers. The other party, which often agrees with the benefits of nutrient recycling, is concerned about the associated recycling of hazardous substances to arable land, possibly endangering long-term soil fertility and jeopardizing the trust in crops grown on the fields.

The underlying problem is the metabolism of a society which is today strongly influenced by previously nonexistent material flows: fossil fuel, traffic, products of chemical and pharmaceutical industries, an intensive food industry, and heavy metals in buildings and vehicles, for example (Kroiss 2004). The complex metabolism of materials and hazardous substances in society is an obvious challenge for wastewater management, which can be expected to require new approaches.

The flow of hazardous substances from society to the environment, which is a consequence of industrialisation, urbanisation, and welfare, is built into the society's physical infrastructure as well as our social behaviour. Consumption is an important factor in these ongoing diffuse emissions. Consumption was recognised as one of the most central and also one of the most disregarded elements in the global search for sustainable development by the Worldwatch Institute (2004). Consumption is basic to human well-being, but consuming too much or consuming the wrong things undermines both our personal health and the health of the natural environment on which we depend (Worldwatch Institute 2004).

New consumption patterns will be required to lift billions of people out of poverty in a manner that is consistent with global sustainability, which implies consumption being in part a societal challenge that will require effective use of regulation by the government and instruments of financial policy, to achieve the common good (Worldwatch Institute 2004). Nevertheless, we all make individual daily life decisions (which from time to time are well thought-out but more often unawares) that affect not only our own communities, but also the world as a whole, both its current and its future inhabitants (Worldwatch Institute 2004). Andrén *et al.* (2004) presented a rather pessimistic analysis of the driving forces for today's and future consumption:

"In consideration of the many different factors that influence modern people's choices of lifestyle and consumption patterns, it is hard to believe that information and rational arguments alone noticeably will affect these choices. Modern society offers each individual an existential vacuum and the imperative to fill it with meanings; as long as consumption remains the primary strategy to do this, purchasing power and resource depletion will go hand in hand."

In view of consumption and the subsequent diffuse emissions, what visions are realistic for the management of urban water and wastewater regarding the flow of hazardous substances in wastewater systems? Seemingly, the phenomenon is partly driven by players outside the urban water and wastewater sector. Accordingly, the flows of hazardous substances in wastewater systems are not only a complex issue for wastewater management, but for society as a whole.

Terminology in this section

Substances: chemical elements and their compounds in the natural state or obtained by any production process.

*Hazardous properties***:** the inherent capacity of a substance to cause adverse effects.

An important distinction is the difference between the terms *hazard* and *risk.* *Hazardous* refers to the inherent properties of a substance, while the *risk* indicates the probability that adverse effects may occur in a human population or environmental compartment due to exposure to a substance. Accordingly, there will be no risk without any exposure, or vice-versa: a very toxic (hazardous) substance may already pose a risk at low exposure, while a higher exposure is required for a less toxic substance to cause a risk (European Commission 2001a).

Box 3-1. Terminology

3.3.1 Tools for management of hazardous substances in wastewater systems

Systems for sustainable urban water services require tools to assess and control various risks related to urban water and wastewater flows. Existing waterborne sanitary systems signal to their users the removal of their (unwanted) waste by just opening the tap or flushing the toilet. Therefore, it is relevant to search for wastewater management tools that support a shift in perspective by combining a traditional end-of-pipe viewpoint with more system oriented ones, thereby linking the use of resources and the spreading of hazardous substances to their underlying causes and driving forces (*i.e.* consumption and lifestyle) rather than focusing only on the emissions.

The flows and sources of wastewater fractions and their constituents, such as nutrients and hazardous substances, are essential knowledge when assessing optional wastewater strategies. To this end, substance flow analysis (SFA) is a useful tool that provides the base of the barrier approach described below.

The SFA methodology was applied to two scenarios to track the flows of hazardous substances in wastewater systems. The 17 hazardous substances selected (Ag, Cd, Cr, Cu, Hg, Ni, Pb, Pt, Sb, Sn, Zn, 4-Nonylphenol (4-NP), Anthracene, Benzo(a)-pyrene, Di-(2-ethylhexyl) phtalate (DEHP), Penta brome diphenyl ether (PentaBDE), and Triclosan) were assessed in a comparative SFA of a conventional scenario vs. a separating scenario, see Figure 3-10. The comparative SFA revealed the conventional scenario to cause an overall higher flow of the selected hazardous substances to the receiving environment, *i.e.* the receiving water and arable land, than the separating scenario. In the separating scenario, parts of the hazardous flow were directed to the landfill, while the bulk emerged in the greywater. Exceptions were Ag and Sn, which subsisted to about 80% in blackwater, and Hg, Zn, and 4-NP, 40-60% of which occurred in the blackwater. Of the remaining 12 substances, only a minor part up to 20% remained in the blackwater. The comparative SFA studies are described in further detail by Palmquist (2004a) and by Malmqvist and Palmquist (2005).

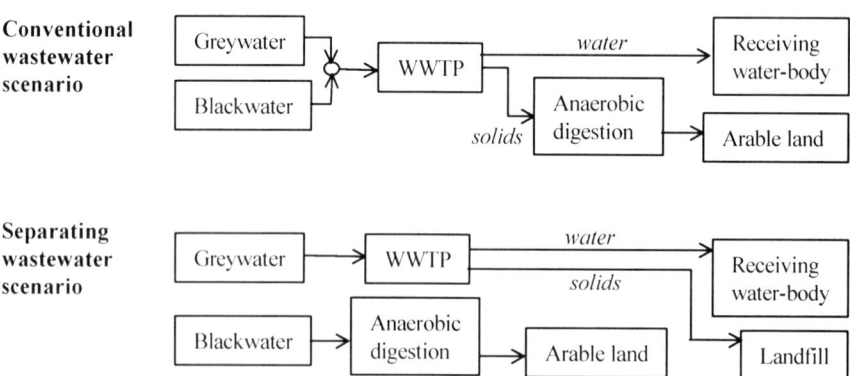

Figure 3-10. System boundaries for, and the design of, two wastewater management scenarios.

3.3.2 A barriers approach

A barriers approach for the assessment of hazardous flows in municipal wastewater systems was suggested (Palmquist 2004a). The Oxford English Dictionary (2004) defines a barrier as "a fence or material obstruction of any kind erected (or serving) to bar the advance of persons or things, or to prevent access to a place". In wastewater management, the intended use for the barriers approach was to interpret and to compare different wastewater systems, and also find out whether and how much the flow of hazardous substances can be stopped, diverged, or transformed at the source or during transport throughout the system. Five kinds of barriers were suggested (see Figure 3-11).

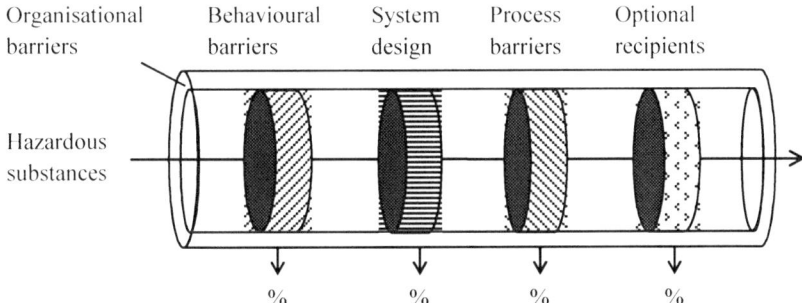

Figure 3-11. A schematic outline of the barriers concept illustrating four of the suggested barriers embraced by the organisational barrier (the tube), which by legislative and administrative measures can affect directly or indirectly all of the other barriers. Each type of barrier can potentially reduce the hazardous flow by specific percentages (here illustrated by %).

Organisational Barriers

Organisational barriers with the aim to prevent risks from the flow of hazardous substances in society represent a wide spectrum of legislative and administrative measures at global, national, and local levels. In Europe, general large scale regulatory and organisational changes are preferably governed by the EU; for example, in the field of water policy, the Water Framework Directive (WFD) specifies objectives for achieving a good status for all European waters by 2015, with sustainable water use throughout Europe (EU Water Framework Directive 2000).

The WFD will affect the institutional arrangements and incentives of the urban water and wastewater sector within all EU member states, including the monitoring and management of hazardous substances in wastewater. Nationally, central governments may establish various kinds of organisational barriers for hazardous flows in society as a part of their ambitions to protect human health and the natural environment.

Potential organisational barriers for hazardous flows in society are:

- Chemicals policy, *e.g.* REACH (Registration, Evaluation and Authorisation of CHemicals) based on the "White Paper – Strategy for a future chemicals policy" (European Commission 2001a);
- Emission regulation, *e.g.* the IPPC directive (Integrated Pollution Prevention and Control) (Council Directive 96/91/EC);
- Technical regulations for the design and function of wastewater systems;
- Fertilising policies, *e.g.* the approach of the farming and food industries to the use of wastewater residues (*e.g.* sewage sludge) on arable land; and
- Eco-labelling regulation (Regulations European Commission No. 1980/2000).

Phasing out or substituting hazardous substances is a possible legal measure and an example of implementing an organisational barrier. For example, the current objective is to heavily regulate or even phase out the use of cadmium. There are however two sides to this. On one hand, cadmium is mined as a by-product of zinc; if the OECD (Organisation for Economic Co-operation and Development) countries phase out cadmium, the price will drop, possibly resulting in uncontrolled use in non-OECD countries. On the other hand, if cadmium is extensively used, for example in solar cells made of cadmium telluride, or in nickel-cadmium batteries, or both, its value would increase and more wasteful applications would decrease. Referred to as a soak-up strategy, this could also be applied to other flows. The strategy could provide an incentive to reduce the leakage of toxic metals to the environment (Azar *et al.* 2002).

Bans on detergents containing phosphate are another example of the phase-out strategy. Banning was successful in reducing the phosphate flows to the receiving water bodies, but as the tenside compounds that often replaced the phosphate turned out to be persistent (and thus relatively resistant to degradation in the WWTPs), the phasing out of phosphate (done legally, *i.e.* an organisational barrier) replaced one environmental hazard with another. These examples highlight the importance of studying how material flows are nested. For this reason, the aim should not always be to phase out specific substances, but to phase out their use in specific applications.

Behavioural Barriers

Behavioural barriers address the consumer perspective. What products are currently used? What barriers for hazardous flows are to be set up in the households? What effects on hazardous flows can be expected of information campaigns? A straightforward example is the information campaign, about cadmium in artists' paint, conducted by the Stockholm Water Company. Artists' paint may contain up to 45% Cd, which is the pigment in these paints. According to the Stockholm Water Company, their municipal WWTPs receive more than 30 kg of Cd per year originating from these paints. They recommend the use of alternative paints, and instruct how to handle the cleaning of brushes and waste. The barrier effect of such measures is, however, very difficult to assess, and one should probably not be overconfident in the response. As a consequence, it becomes essential to phase out or replace hazardous substances in consumer goods and products.

System Barriers

System barriers relate to the infrastructural and technical design of urban water and wastewater systems. Examples are the separation of urine, faeces, greywater, and stormwater at the source, while combined flows that mix wastewater originate from numerous other sources. In our own study, one quarter of domestic wastewater phosphorus emerged in the greywater and three quarters in the blackwater (Malmqvist and Palmquist 2005). For cadmium and triclosan, the result was almost the opposite, *i.e.* 80% in the greywater and 20% in the blackwater. This type of relevant information about a system barrier is needed to decide on the design of a technical system.

Technical or Process Barriers

Technical or process barriers involve the process units in wastewater treatment plants. These are barriers, based on mechanical, biological, and chemical treatment processes, which achieve separation, decomposition, or both, of the constituents in the wastewater. However, complete mineralisation of xenobiotic compounds in treatment systems is rare. As presented in Figure 3-12, the biodegradation of xenobiotic substances varies with the operating conditions in the WWTP, and between substances (Palmquist 2004a).

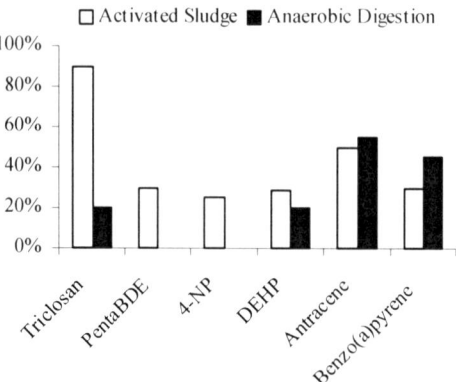

Figure 3-12. The biodegradation of organic hazardous substances by aerobic activated sludge treatment and by anaerobic digestion. For PentaBDE and 4-NP, no biodegradation was expected under anaerobic conditions.

The fate and distribution of hydrophobic chemicals in treatment systems is largely controlled by the physicochemical properties. According to Byrns (2001), biodegradation mostly influences compounds with moderate hydrophobic properties (log K_{ow} values in the range 1.5-4). For biological nutrient removal, long solids retention times (SRTs) are applied. This seems to favour the biodegradation of many, but not all, hydrophobic organic substances (Byrns 2001; Kroiss 2004).

Sedimentation, flocculation, chemical precipitation, sand filtration, and membrane filtration are all separation processes in wastewater treatment. The application of membrane processes (micro, ultra and nanofiltration, as well as reverse osmosis) in the treatment of water and wastewater is increasing. A membrane separates wastewater into two streams, a purified one and a concentrated one that needs to be taken care of.

In Malmqvist and Palmquist (2005), a nanofiltration (NF) membrane was modelled as an additional process barrier in the WWTP; this considerably reduced the substance flow to the receiving water body in both the conventional and separating scenarios (see Table 3-5). Data for the modelling was gathered from Visvanathan and Roy (1997), who reported a phosphorus separation efficiency for NF as tertiary treatment of wastewater to be >95%, resulting in an effluent concentration of less than 0.1 mg P/l.

The NF process reduced Cd^{2+} from 500 mg/l to 15 mg/l, corresponding to a 97% separation efficiency, as reported by Qdais and Moussa (2004). The NF separation of triclosan was assumed in our work to be 80%. The principal disadvantages of membrane filtration in wastewater treatment are higher costs and the operation and maintenance requirements as compared with conventional treatment methods. Applying membrane filtration directly to wastewater may often be problematic, due to the variable composition and high fouling potential of most wastewater (Côté and Thompson 2000).

Table 3-5. (A): mass flow in the effluent from the WWTP after mechanical, biological and chemical treatment. (B) the same as in (A), plus nanofiltration (Malmqvist and Palmquist 2005).

Substance	unit	Municipal wastewater		Separated greywater	
		(A)	(B)	(A)	(B)
P	kg / year	221	11	57	3
Cd	g / year	76	2.6	60	2.2
Triclosan	g / year	37	7	30	6

The separation of pharmaceuticals from nutrients in human urine by nanofiltration was investigated in a laboratory study as an example of a process barrier (Palmquist 2004b). Several nanofiltration membranes were tested for the separation of pharmaceutical and estrogenic compounds from urine, to generate a micropollutant-free nutrient solution to be used as fertiliser.

A fresh urine solution containing most of the nitrogen in the form of urea, and a synthetic one with a similar inorganic composition, were tested at various pH values to investigate the separation behaviour. These solutions were spiked with the pharmaceutical and estrogenic compounds propranolol, ethinylestradiol, ibuprofen, diclofenac, and in some cases, carbamazepine. The retention of both pharmaceuticals and inorganic ions was influenced by pH. In general, with increasing pH the acidic compounds (ibuprofen and diclofenac) exhibited higher retention, while the basic (propranolol) and neutral (ethinylestradiol) compounds had a lower retention.

Among the membranes tested, the NF270 membrane showed the best performance in retaining the pharmaceutical compounds (Figure 3-13). Optimum retention of the pharmaceutical compounds was obtained at pH values around 5. At this point, the retention of all the pharmaceuticals in human urine was above 92%, while the retention in the synthetic urine solution was above 75%. The difference in retention could be partly explained by the influence of organic matrix substances in human urine. These substances (such as oxalic acid, uric acid, amino acids, and the like) can form complexes with Ca^{2+} and Mg^{2+}.

Retention

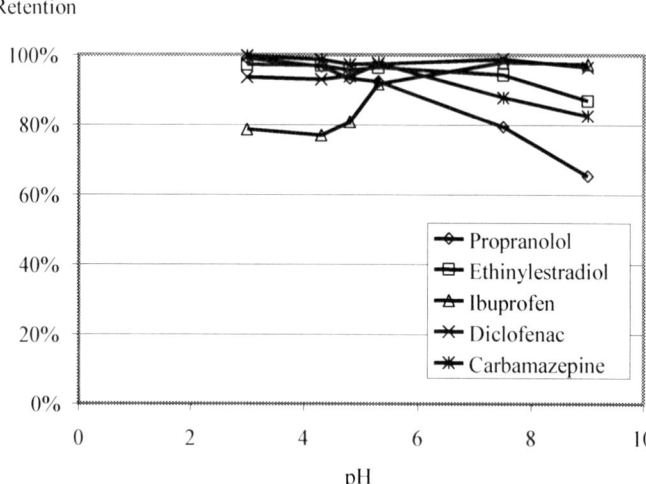

Figure 3-13. Retention of pharmaceuticals as a function of pH in fresh urine using an NF 270 nanofiltration membrane.

Organic compounds or their complexes can adsorb, essentially, to the membrane by electrostatic or unspecific interactions (van der Waals), perhaps functioning as a secondary membrane, and thereby increasing the retention of the organic compounds.

The NF separation of ions is largely based on electrostatic interactions, as displayed by the much higher retention of multivalent ions (phosphate and sulphate) than of univalent ions (Na^+, K^+, Cl^-, $NH4^+$) (Figure 3-14). Furthermore, non-charged compounds such as urea showed much less rejection, which means that the multivalent ions (phosphate and sulphate) were retained together with the pharmaceutical compounds in the concentrate, The urea passed the membrane, thus forming a nitrogen rich solution largely free from pharmaceuticals.

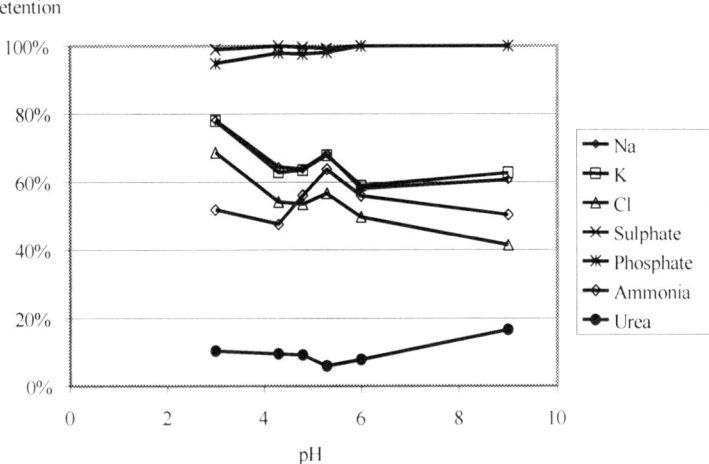

Figure 3-14. Retention of inorganic ions as a function of pH in fresh urine using an NF270 nanofiltration membrane.

The results from the study showed that membrane filtration is a potentially promising process barrier. Membrane processes however need thorough optimisation, *i.e.* inaccurate operating conditions and operating failures may result in rapid deterioration of the membrane.

Optional Recipients

Assessing and choosing optional recipients is a way to protect sensitive recipients. By selecting another recipient for wastewater residues, sensitive recipients may be protected by redirecting the hazardous flow elsewhere to a less sensitive recipient. Optional recipients, which could be lakes, rivers, seas, or soils, are highly dependent on the geographical context. As was shown in the SFA studies, the barrier effect for Cd was only moderate in both the conventional and the separating wastewater management scenarios (Palmquist 2004a; Malmqvist and Palmquist 2005). An additional NF membrane was shown to protect the receiving water body; however, to protect the arable land, additional measures would be required. Here, optional recipients could be a matter of discussion.

As stated in Palmquist and Jönsson (2003), the fertilising potential of wastewater sludge must be questioned in a long term perspective, since the metal to nitrogen ratios of 12 hazardous metals (including Cd) revealed ratios in sludge that are higher than what the plant uptake can counterbalance, thus implying metal accumulation in the soils. This may harm *e.g.* the soil fertility and the production capacity of the soils (Palmquist and Jönsson 2003). Due to this risk, the sludge from the combined wastewater system might be ᵤₛₑᵈ ... other soil applications than as fertiliser for food production, to safeguard clean food production. LeBlanc *et al.* (2004) propose various ways to use wastewater sludge: composting, mine site reclamation, landfill cover, tree farming, and topsoil manufacturing.

The combined barrier effect

It is important not only to identify each barrier within a defined system, but also to assess the combined effects from all existing barriers. Evaluating the combined barrier effect implies considering the whole chain of barriers in reverse order: from the actual receivers of the hazardous substances, *e.g.* the receiving water, arable land, and landfill, to the sources. To obtain the combined barrier effect in the system, the amount by which each substance is reduced at each barrier is multiplied. Figure 3-15 shows the combined barrier effects (counting systems and process barriers) for the case of the small town of Surahammar (see also Section 9.2) for (A) emissions to water and (B) emissions to arable land (Palmquist 2004a).

An evaluation of the combined barrier effect implied that a change from a conventional wastewater system to a source separating system would have a greater impact on the management of solid residues (*i.e.* the emissions to arable land) than on the effects in the receiving waters (Figure 3-15). The flow of hazardous substances to the receiving water would not be greatly affected by such a system change, as shown by the small difference in the combined barrier effects for emissions to water. However, high levels of barrier protection do not guarantee chemical safety. Substances passing through the barriers, even in very small amounts, but which are very harmful to the receiving environment, may cause more severe ecotoxicological effects than a high volume of substances of low toxicity.

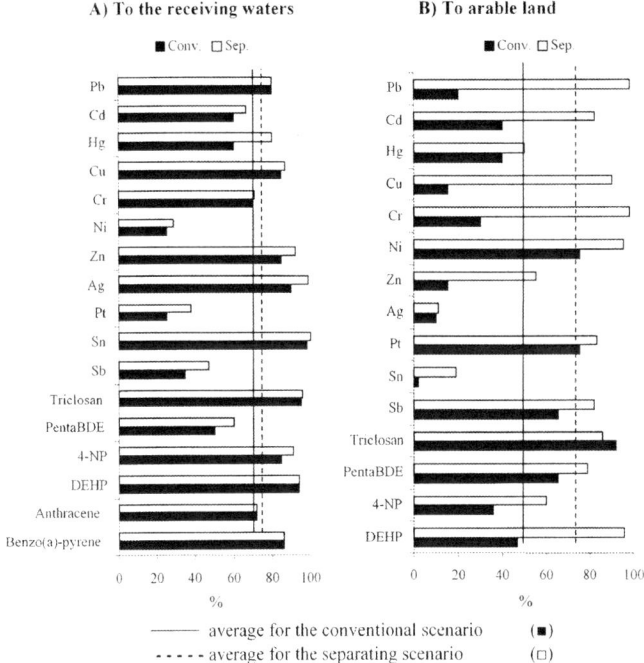

Figure 3-15. The system and process barriers' combined reduction (%) of the pollutant transport to the final recipients: (A) the receiving water, and (B), the arable land. The barrier effects for emissions to arable land could not be evaluated for the two PAHs due to a lack of data.

The barriers approach was proposed as a tool on a conceptual level (a way of thinking), as an attempt to generate a basis for systems analysis, risk assessment, improvements of the system, and to facilitate communication about hazardous substances in the wastewater systems.

Barriers depend on the context, implying that a barriers model developed for the wastewater system in one municipality may not be fully applicable to the wastewater system in another municipality due to such differences as physical structures, the inherited infrastructure, the number of citizens, and environmental ambitions.

Furthermore, the different kinds of barriers need to be worked out and managed on several levels. For instance, organisational barriers require measures that start with national and international legislation, while system barriers are decided upon and managed by urban planners and engineers at the regional and municipal levels. Behavioural barriers are influenced by individual choices and by the range of products that is on the market. Therefore I would, again, like to stress that the flows of hazardous substances in wastewater systems are a very complex issue, not only for wastewater management, but also for society as a whole. Several kinds of measures are needed to prevent the spreading of hazardous chemicals to the environment.

4

Assessment: economy

4.1. Bo Olin
4.2. Marianne Löwgren

4.1 BUSINESS ECONOMICS

The Urban Water Toolbox contains tools for strategic choices. Strategic choices may involve policy changes, long-term or global system changes, or new heavy investments. Very often, the decision is made on a higher level than the level of the actual system. In the context of urban water systems, this means that many strategic decisions are made in connection with long-term municipal or regional planning, or planning according to the Water Framework Directive. Of course, many strategic decisions are taken in the planning processes within the water administrations and the real estate companies.

Urban water systems have very long life expectancy, both technically and economically. The life expectancy of system components can vary from 5 - 10 years and up to more than 100 years. The economic life of a system is a flexible concept depending on how the system is maintained and renewed or new investments are made. Reinvestment in this context means steps taken to retain the capacity of the system. The life of a system, both technically and economically, can also be prolonged by improving existing processes and adding new ones within the framework of the present system. This is an ongoing business mostly activated by technical development and environmental legislation. In this way, the initial large investments are kept alive for longer and longer periods, and new investments are added within the old framework.

In common with many other infrastructural systems, urban water systems are characterized by economies of scale, thus a larger system has lower investment and operating costs per m^3 water and wastewater than a smaller system. Although this is not a natural law, for infrastructural systems this maxim seems to be true within a wide context.

The discussion above indicates that a costing model must be simple, immediate and user friendly. At the same time, the model must be able to discriminate between alternative structures and built up in a way that offers simple sensitivity analysis. The traditional detailed cost estimation is too time consuming, costly and complex. A top-down approach, where one calculates with aggregate parameters seems to be preferable. The top-down approach does not require the detail of the bottom-up approach and enables the analyst to estimate total costs with less effort. The analyst may relate activities within alternative system structures to similar activities with known costs. The applied data are empirical and derived from earlier experience (Roy 2003). For the strategic choice situation in our context, the top-down approach appears preferable.

4.1.1 Background

The value contained in wastewater, though relevant from a sustainability point of view, has a low market acceptance and thus a low market value. The income potential in an urban water system is consequently difficult to realize in a short run. These conditions can change gradually due to higher prices on energy and fertilizers. Urban water systems are characterized by heavy investments, long life expectancies and economies of scale. This means that the systems are resistant to radical changes. What types of cost calculation do the decision makers need in such an environment? Table 4-1 illustrates several types of cost calculations.

The table indicates that the greatest need for calculations applies to marginal changes within existing systems. One has to decide how new residential or industrial areas will be served with water and wastewater systems. How bottlenecks in the processes are to be remedied and how to react to technical innovations, *e.g.* new processes, must also be decided. In addition, there are

decisions about living up to new environmental restrictions. In Sweden, new
urban water systems will be very rare for many years to come. For example in
developing countries, the situation can be quite different.

Table 4-1. Several types of cost calculations for urban water systems.

Type of calculation	Example	Relevance
New investment, total system	Totally new system	Very rare*
New investment, large part of	Relocation of sewage plant	Rare
system	New sewerage for stormwater	Occasionally
New investment, small part of	New process	Frequent
system	New area, pipelines	Frequent
	New area, processes	Occasionally
Expansion	New area	Frequent
	Process	Frequent
Improvement	Renewal	Frequent
	Increased capacity	Frequent
Reinvestment	Relining of pipelines	Occasionally
	Change of main pipelines in buildings	Frequent
Maintenance	Retain capacity	Frequent

* in Western Europe and North America

Roy (2003) gives an overview of some estimation methods that can be used in
an early phase of a development process. His references are mostly within
product design in the aircraft industry and software development, but he
generally talks about cost estimation in an early development phase where many
choices are still open. The conditions agree quite well with those strategic
decisions that we envisage for the Urban Water Toolbox. Roy describes five
methods of early phase estimations.

- Parametric Estimation (PE)
- Neural Network (NN) based cost estimation
- Case Based Reasoning (CBR)
- Feature Based Costing (FBC)
- Detailed Cost Estimation (DCE), a first sight estimate that is done early

Table 4-2 shows the stages of development at which the five methods are
applicable.

Table 4-2. Cost estimating techniques and product lifecycle. Source: Roy (2003).

Used for	Cost Estimation Techniques				
	PE	NN	CBR	FBC	DCE
Concept design phase (innovation)	X		X	X	
Concept design (similar products)	X	X	X	X	
Feasibility Studies	X	X	X	X	
Project definition	X	X	X	X	
Full Scale development				X	X
Production				X	X

The two methods most interesting for our aims are Parametric Estimation (PE) and Case Based Reasoning (CBR). The CBR is based on building up a database that contains data from many individual case studies. The CBR is a search engine for finding analogous cost data in an actual case. Parametric estimation is described as a widely used method for estimating product cost at the early stages of a project by using a Cost Estimating Relationship (CER). While CBR could be interesting for the future if a case database is constructed, PE is currently the more convenient method to use because the user can build up own CERs.

Cost estimating relationships are the nuclei of the parametric model (Brundick 1995). A CER consists of a value assigned to a cost driver and a statistical function. The cost driver is the independent variable: it can be an inflow to or an outflow from an urban water system or component. The statistical function can be a linear or non-linear equation that includes the value of the cost driver. Some equations can be the result of regression analysis; other equations can be very simple. In this way, cost parameters can be constructed if there are historical data from similar processes or components.

When there is no historical data, a cost relation can be generated in two ways. Analogy means that one estimates the costs for an unknown entity by comparing the unknown to a known similar entity and its cost. If historical data for a digestion plant is lacking, but the process is believed to be similar in cost to an incineration plant, the historical data for the latter can be used. It is furthermore possible to estimate the cost of a new process from a prototype. However, it must be kept in mind that with a prototype cost for many processes is scale dependent.

4.1.2 Tool description

The Urban Water costing tool consists of a parametric costing model built in Excel 2003, which contains empirical cost parameters, historical cost parameters and statistical equations for investments and cost of operations for some predefined system components. The model has the capacity to compare up to six different system structures. For each component or sub-component the user has to fill in certain base line data (cost drivers), which trigger the cost calculations. Users can also add their own cost data for system components that are not predefined in the model. The model uses Swedish language and currency, and has been developed to match costs for urban water systems under Swedish conditions. With some effort, the model can be adapted to other countries.

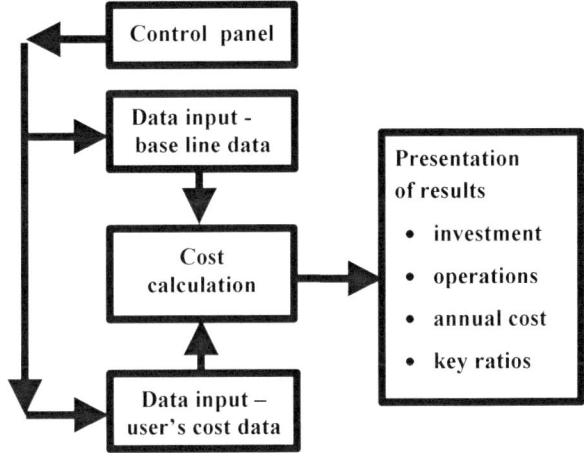

Figure 4-1. Structure of the parametric costing model.

The model (Figure 4-1) consists of the following parts:

1. A control panel from which the user can navigate with the help of keys;
2. A user interface for base line data input
    a. description of the object and the system structures to be compared;
    b. common base line data for all system structures;
    c. specific base line data for each system structure;
3. A user interface for cost data input of the user's own system components;
4. A cost calculation containing predefined system components with CERs for initial investments and annual costs of operation and maintenance (this part is available to the user but does not allow changes); and
5. A presentation of absolute and relative total annual costs and key ratios.

The model is flexible to changes. It is easy to add or remove components, and to make a sensitivity analysis by changing the input data.

4.1.3 Data need

The cost drivers for the parametric costing model are shown in Figure 4-2.

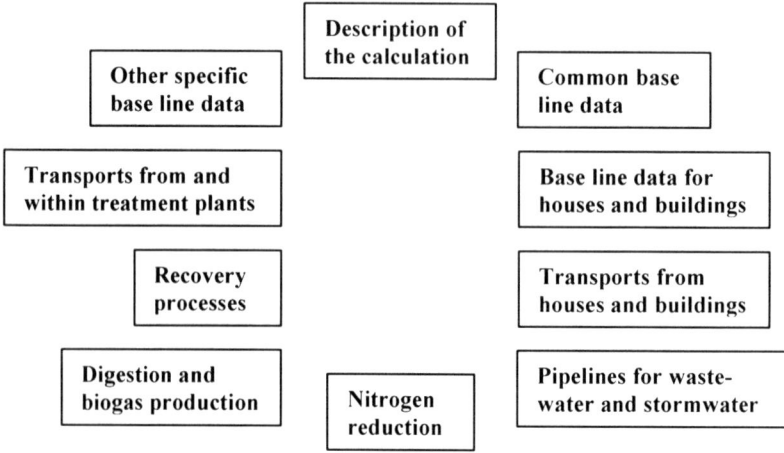

Figure 4-2. The types of base line data input shown on the control panel. Nitrogen reduction is an add-on process in the model and therefore has its own button.

Below, the ten types of input data are listed.

- Description of the calculation: The user describes the aim of the calculation, the parameters, system structures to be compared, and the parameters for depreciation and interest rate.
- Common base line data: Data that applies to all system structures, for example number of persons connected. It is also possible to enter input data for existing plants and processes.
- Base line data for houses and buildings: The user describes installations in houses and buildings for the system structures. Here you can specify the percentage of specific types of installations in the area of investigation.
- Transports from houses and buildings.
- Pipelines for wastewater and stormwater.
- Nitrogen reduction.
- Digestion and biogas production.

- Recovery processes.
- Transports from and within treatment plants.
- Other specific base line data: refers to the calculations for a conventional wastewater treatment plant, incineration, landfill and storage of urine.

Users can also define their own system components. When this is done, they also have to register their own cost data for investment and operations. Table 4-3 shows an example of an input form from a case study. This screen can be reached by choosing Recovery processes on the control panel (Figure 4-2). In this example, four system structures including alternatives for sludge handling were compared. These comprised 1) the existing system with soil production; 2) the existing system with urine separation; 3) direct use of sludge as fertiliser, and 4) AquaReci, a process for nutrient recovery from sludge. Six system components were predefined in the model. In this example, the user has registered volumes and quantities for each structure and component where relevant.

Table 4-3. Input form for recovery processes. Example from a case study that included the AquaReci nutrient recovery process.

Component		System structure			
		1 Soil	2 Urine	3 Sludge	4 AquaReci
RO[a]-process	m^3/d	0	0	0	0
UASB[b]-reactor	m^3/d	0	0	0	0
AquaReci[c] (TS 15%)	t TS/year	0	0	0	22868
KREPRO[d]-plant	t/year	0	0	0	0
Composting of sludge or other organic waste	t TS/year	3560	3560	0	0
Soil production	t/year	5371	4922	0	0

[a] RO = reverse osmosis; [b] UASB = Upflow Anaerobic Sludge Blanket; [c] AquaReci = Process based on supercritical water oxidation (see e.g. Gidner et al. 2000); [d] Kemwater Recycling Process: thermal hydrolysis at low pH (see e.g. Eliasson et al. 2000)

4.1.4 Using the tool

The cost calculation database

In the model, there is space for six system structures to be compared. Every system structure is the sum of predefined system components with CERs for initial investments and annual costs of operation and maintenance. The user may add new components together with cost data, and the calculations are triggered by the input data. The results are stored in the cost database as initial investment costs and annual costs of operations and maintenance for each system component including the user's own components. Table 4-4 shows the main predefined system components.

Table 4-4. Predefined system components in the main calculation. Some of these components have predefined sub-components.

Component	Unit
Houses and buildings	From sub-model
Transports from houses and buildings	From sub-model
Pipelines for wastewater and stormwater	From sub-model
Pumping stations	From sub-model
Stormwater facilities	From sub-model
Conventional treatment plant (1)	m^3 / d
Nitrogen reduction	From sub-model
Digestion and sludge handling	From sub-model
Recovery processes	From sub-model
Storage of urine	From sub-model
Incineration	Tonnes / year
Landfill	Tonnes / year
Transports from plants and processes	From sub-model

Example of a CER: The conventional wastewater treatment plant

The treatment process includes mechanical, biological and chemical treatment. Nitrogen reduction is not included. The CER for investment is a non-linear function $C = 579\ 150\ K^{0.55}$, where K = wastewater volume (m^3/d) and C = Total investment cost in Swedish crowns (SEK). The Excel formula contains two alternative calculations: 1) new investment, 2) expansion. For the first, the model calculates the investment from the total volume. For the second, the difference between the investment cost for the volume after expansion and the volume before expansion (existing plant) is calculated.

4.1.5 Presentation of results

The data in the cost database are computed to obtain final results of the comparison between the system structures. These results comprise annual costs per system component and group of components in both absolute and relative terms. The total annual cost consists of the annual capital cost and the annual cost for operations and maintenance. Annual capital cost is calculated from initial investment, depreciation time and interest rate.

The system components are sorted in four groups: Houses and buildings, distribution (transports of rest products from estates, and sewerage), plants and processes, and transports from plants and processes. The presentation is structured in three parts:

Part 1. Annual capital cost, annual cost of operations and total annual cost are presented per system component and in total for each system structure.

Part 2. Total annual cost is given per component group and in total for each system structure in a table and in a diagram. In the table, the transition costs from a reference structure are shown (Table 4-5).

Part 3. The following key ratios are given per component group and in total for each system structure in a table and in a diagram:

- Total annual cost per connected person,
- Total annual cost per cubic metre of wastewater,
- Total annual cost per recovered quantity of P, K, N and S (if available),
- Total annual cost per discharged quantity of P and N (if available), and
- Total annual cost per kWh energy used (if available).

Table 4-5. Example of presentation table for consolidated annual costs per component group for each system structure. The Table also shows the transition cost from the reference case "1 Soil" to the other alternatives. Annual costs (10^3 SEK)

	System Structure			
Component	1 Soil	2 Urine	3 Sludge	4 Aqua Reci
Houses and Buildings	160 112	166 541	160 112	160 112
Distribution	24 392	24 763	24 392	24 392
Plants and processes	86 329	87 666	72 501	85 469
Transports	139	139	139	38
Sum	270 973	279 109	257 144	270 012
Transition cost		8 136	-13 829	-961

4.1.6 Discussion

Since each study has its own focus and extent, every comparison of systems is unique. This means that system components in play are not the same and that it is difficult to compare the results of studies made on different municipalities. It also means that the users need a smorgasbord of system components when designing their system structures. Local variations are common and it is hard to predefine system components for all of them.

When evaluating long-term solutions, some of the alternative structures are likely to contain new processes for which there are few historical data. In the Urban Water model cities, new processes such as AquaReci and some blackwater systems have been tested. In these cases, costs from prototype plants were used despite the fact that the cost of small-scale plants is no good estimator for the cost of full-scale processes. New Cost Estimating Relationships (CERs) must be developed as soon as full-scale cost data are available.

A parametric costing model was developed that has both limitations and potential. It is viable and ready for application, and using it under real conditions is the best test of a model. It was developed in Sweden in collaboration with Urban Water model cities. However, we think that the model can also be used in countries where the infrastructure in urban areas is less developed. We have pointed out some areas in which further research and development can raise the quality of the model.

4.1.7 Model status and availability

Many of the CERs in the model are still preliminary. They are based on limited, and historical data that is either outdated or new and not yet confirmed. To survey and construct cost estimation relationships for specific system configurations is an interesting task. Let us take some digestion processes as an example. Here we can discern applications such as wastewater sludge digestion, blackwater digestion (either with pre-sedimentation or not dewatered), organic waste digestion, greywater sludge digestion, or combinations of these. We can add a co-digestion plant for slaughterhouse waste, waste from farms etc. All these combinations can be very worthwhile alternatives when the municipalities make their strategic choices in the future. (Svärd and Jansen 2003; Jansen *et al.* 2004). To map all these processes and calculate CERs, perhaps together with a model designer, is an essential task.

The aspects of income have not been given justice yet in the model. We aimed originally to include all commercial income and cost reductions within the system boundaries. As we understand, there are many income potentials. Unfortunately, the market coupling is rudimentary and the market acceptance is low. Most of the incomes consist of internal cost reductions. In this situation, we decided to exclude the income side in this version of the model. Since it is technically feasible to include them, they should be integrated in due course. However, we believe there is a need for more research. How do we calculate income for the future, for example 30 to 50 years from now? How do we handle future energy prices and the approaching lack of non-renewable resources, such as phosphorus?

Detailed documentation on the Excel model can be found in Olin *et al.* (2005). The model and the documentation are in Swedish language. Further information in English is available through contact with the editors.

4.2 SOCIETAL AND ENVIRONMENTAL ECONOMICS

The purpose of the calculation model elaborated in the preceding chapter is to guide decision making for water utilities and municipal planners. Business economics concentrates on business opportunities and financial costs. There are however wider considerations that should be taken into account to reach the comprehensive goal of sustainable urban water management, which include environmental and social aspects.

4.2.1 Full economic costs

While financial costs represent the perspective of business entities and municipalities, *full economic costs* are based on a societal perspective. Full economic costs are calculated to facilitate cost-effective decisions and cost-recovery for sustainable water services. They should include three components: financial, resource, and environmental and social costs (Table 4-6).

Table 4-6. Definition of the full economic costs.

Components of full economic costs	Examples
Financial costs (including internalised environmental costs and transfers)	Operation and maintenance, capital costs, administration costs
Scarcity costs	Costs for limited resources
Environmental costs	Pollution, environmental damage
Social costs	Reduced welfare

When calculating financial costs, subsidies and transfers must be separated to correct for non-economic costs. Some environmental transfers, *e.g.* environmental taxes and state grants for building wastewater treatment plants, may be internalised environmental costs. If water resources are scarce, market prices should be compared to the opportunity costs[1], to account for resource costs. Scarcity means that there are competing uses that might outweigh the benefits of a specific use. Some of the environmental costs are water related, although environmental damage also occurs in other compartments. Many environmental resources cannot be assigned explicit costs because there is no market for them.

---

[1] Scarce natural and human resources can be used in various ways, some of which are more profitable than other. The opportunity cost of any factor unit in a particular production is the maximum amount that the factor unit could earn in some alternative use.

Minimum requirements for the presentation of cost information (EEA 1999)

1. It is essential that reported costs are properly defined. As a minimum, the total investment expenditure and total annual operation and maintenance costs should be reported separately.

2. As far as possible, it is recommended that all cost data should be documented in full in the year in which the actual expenditure is incurred, even if the data are subsequently adjusted to take account of time, for example, by using discount rates.

3. All costs should be measured in relation to an alternative. The alternative most commonly employed is a projection of the existing situation, *i.e.* the situation in which the environmental protection measure has not been installed. Therefore, only additional costs actually incurred relative to the base case should be included in the reported cost data.

4. Where the costs associated with an environmental protection measure have been apportioned between two or more controlled pollutants, the method of apportionment should be described.

5. The reported cost data should only related to direct costs; indirect costs should be excluded.

6. Where environmental protection measures produce non-environmental benefits, revenues or avoided costs, these should be reported separately from investment expenditures and operating and maintenance costs.

7. It should be remembered that costs and prices are not fixed. For example, the unit price of a measure often falls as it changes from an experimental measure to a mass-produced one. Therefore it is recommended to use the most recent valid data available.

8. It should be remembered that old equipment can sometimes have a lower efficiency and higher maintenance costs than new equipment.

9. As a minimum, any discount rate used should be recorded.

10. If cost data are adjusted for inflation or changes in price through time, then the method used should be recorded and any index used should be recorded and referenced.

11. If determining annual cost data, the approach which has been used to derive the annual costs should be recorded, along with all underlying assumptions.

Box 4-1. Minimum requirements for the presentation of cost information. source: European Environmental Agency (EEA 1999).

4.2.2 Cost benefit analysis

Cost benefit analysis (CBA) aims to determine the net benefit to an entire community of an activity, a project, or a policy. The aim of CBA is to balance costs and revenues, both financial items, and positive and negative externalities, *i.e.* unintended effects on the environment or on the welfare of people outside the scope the calculation. In the water business, negative external effects are more common than positive ones. CBA derives directly from the welfare theory in which the value of an activity or a project is measured by the additional benefits to all consumers, weighed against the disadvantages. The foremost purpose of a CBA is to provide a quantitative basis for decisions, with costs and benefits that are expressed in monetary terms.

To conduct a CBA, Hanley and Spash (1993) suggested an analytic structure that should cover the following steps.

1. Setting of a goal: What to achieve?
2. Definition of a relevant project. Here, the boundaries of the analysis must be specified. Duration of the project? A zero alternative should be identified as reference.
3. Identification of consumers and other users, who are directly or indirectly affected, positively or negatively, within the boundaries, as given by Step 2.
4. What changes are likely to come from the project, such as improved services, employment, environmental disturbances, and health effects? Are there any displacement effects, *i.e.* elimination of an existing business or facility?
5. What changes are economically relevant, directly or indirectly?
6. Physical quantification of relevant changes, effects, or both, as well as collection of data. What sort of data should be collected, and how can validity and reliability be ensured?
7. Choice of method for valuing changes (benefits and costs). To be comparable the items should be calculated on either an annual basis or discounted to present values.
8. Externalities that cannot be monetarized should be described in words.
9. Calculation of aggregated costs and benefits, to select the best option.

The concept that costs and benefits can be quantified and, after that, the most cost-effective alternative can be adopted, is appealing. However, it is impossible to express all benefits (and costs) in economic terms. The standard argument against CBA is that, when decisions are made about infrastructure, environmental and social effects are valued anyway, although in a less conscious and transparent fashion. Cost benefit analysis should therefore be used with proper consideration to its limitations (Sugden and Williams 1978). In cases where quantitative estimations of all benefits and costs are impossible or too time-consuming, the application of multi-criteria analysis (MCA) is a good complement to economic analysis.

In pure competition, market prices for goods and services reveal the consumers' willingness to pay (WTP); thus market data are available. For environmental goods and losses, where a true market does not exist, there are alternative methods for deriving economic values. Direct methods consist of devising fictional markets to reveal what the buyers are prepared to pay for the asset being evaluated, *i.e.* their WTP, or what payment would be necessary to make them accept some kind of nuisance, *i.e.* their willingness to accept (WTA). Indirect methods are based on observations of existing markets that are related in some way to the asset under evaluation. The relation could be as a substitute (such as bottled water instead of tap water), or as the expenses assigned with a given activity, *e.g.* all costs related to a visit to a fishing camp (Baumol and Oates 1988).

Another approach, benefit transfer, might be an efficient way of using data from previous studies of comparable cases. However, it presupposes that the circumstances and the measures are broadly comparable (Brouwer 2000). Costs and benefits that cannot be given a monetary value should be described in words. Apart from the problems of valuing benefits that are not mediated by the market, the basic hypothesis of pure competition is seldom valid in practice. Thus, prices often reflect neither the marginal cost of production, nor the marginal utility of the goods and services produced.

Benefits are often more difficult to assess than costs, either because of lack of markets for ecosystem services, such as biodiversity, or the ethical concern about the needs of future generations. As an approximation, a cost-based approach can be used. What would it cost to deliver a given type of environmental service, *e.g.* recreational value? Such an estimate might represent a minimum of the desired value. There are various kinds of measures to achieve environmental objectives. In general they are preventive, *i.e.* either avoiding or restoring (Ekins 2000). Nature reserves belong to the first group. Valuable resources are protected from exploitation and degradation.

Other measures are localised to the source of the problem: legislation can be used to stop the use of harmful substances. Furthermore, both source separation of wastes and recycling are means to protect soil and water quality. Rules for manure management and taxes on commercial fertilizers serve the same purpose. For point sources, and also to some extent for the diffuse sources, the polluter pays principle can be applied, as recommended in the Water Framework Directive (EU Water Framework Directive 2000).

On the other hand, when old sins such as abandoned waste deposits or other contaminated sites are discovered, the polluter often cannot be held to account. In many cases, restoration projects of this kind need to be financed by tax money, an expression of a political WTP.

4.2.3 Economists viewpoint on sustainable urban water management

As for to all other spheres of collective action in society, the current planning of the urban water sector should be guided by two basic principles: cost-effectiveness and sustainability. Effectiveness is usually interpreted at *doing things right*. Cost-effectiveness (CE), which provides a ranking of alternative measures based on both their costs and effectiveness, can be presented in two ways, either as costs divided by the effect, or the effect divided by costs. By striving for CE, the best use of available resources is ensured (Siebert 1987).

In his Ph.D. thesis, Hjerpe (2005) shows that the assessment of the economic performance of urban water management depends fundamentally on the analyst's view of the relation between the economic system and the environment. In traditional neo-classic economics, the economic system is separated from the ecological one. Nature serves as a source of raw material supplies. Pollution is treated as negative external effects. Scarcity problems are solved by technical changes and recycling, that are induced by market mechanisms that provide efficient resource allocation. For exhaustible resources, taxes are the means that will make them last longer. The sum of all assets is constant; there is no qualitative difference between natural and man-made capital. Consequently, the substitution possibilities between natural and man-made capital in the analysis are infinite.

At the other extreme, the proponents of *steady-state economics* and *co-evolutionary economics* view the economic system as a subsystem of the ecological system. Unique properties are ascribed to all ecosystems, which means that they cannot be replaced by man-made capital. There are strong limits to economic exploitation: it must not exceed the carrying capacity of the ecosystem. Between these extremes there are theorists who elaborate on the links between economic and ecological systems. Nature supplies inputs to production processes, but nature also offers amenities, and sinks for residual products. The activities of the economic system should be kept within the assimilative capacity of the environment. Exhaustible resources should be replaced by renewable ones. The proponents of *ecological economics* also view the economic system as a subsystem of the ecological one. They widen the analysis to include biodiversity and other environmental services, such as the regeneration of soils, the pollination of crops, the assimilation of waste products, and the purification of water. The carrying capacity of the ecosystem is the limit, and there are also limits to substitution possibilities (Costanza 1991).

The present urban water systems have evolved over a considerable period of time, as part of the whole institutional and physical infrastructure. Incremental improvements will take place provided we succeed in drawing attention to quantitative and qualitative constraints that are becoming generally recognized. By the implementation of new regulations, such as the EU Water Framework Directive, the institutional concept of water resources has been redefined: they are no longer free utilities; they are now scarce goods that should be used in a cost-effective way.

4.2.4 Indicators

Indicators are tools for the analysis of change. They consist of datasets selected to provide information for interpreting changes in the environment, the economy and society. If two or more indicators are combined an index is formed. Indices are usually used at more aggregated analytical levels, such as a regional or national level. Indicator frameworks are a means to structure a set of indicators in a way that facilitates their interpretation. The Urban Water framework defines a core system which includes users and water utilities, while the organisation and the technical structure are separate entities. The following normative principles should guide the process of identifying economic indicators for sustainable urban water management (Segnestam 2003).

- The cost data included in the analysis of a technical structure must not be limited to the public infrastructure; privately owned installations should also be included.
- The method of distributing costs, chosen by an organisation, must be taken into consideration. Water services must be affordable to all consumer groups.
- To choose efficient solutions, an organisation needs reliable information about what resources are required to implement a specific technical structure.
- Externalities need to be evaluated for the urban water system, to enable it to coexist with its social and natural environment.
- Technical systems are limited by the amount of resources (money, time) the users can afford to put into the building and maintenance of a private

technical infrastructure, as well as their contribution to the public technical infrastructure.

- An organisation must be able to obtain financial resources to run the public technical infrastructure, usually by fees collected from users.

There is no universal set of indicators that is applicable in all situations. To select a useful set of indicators, criteria are essential. The following criteria are appropriate to most indicator selections (Hjerpe 2005).

- *Relevance to objectives* means that the indicator selection must be closely linked to the issues involved, and to the targets set by the stakeholders.
- *Clear definitions* should be stated for the indicators selected. They should be distinct, easily understood and communicable.
- *Realistic data collection expense* means that indicators must be practical and realistic; the cost of data collection and analysis needs to be taken into account.

In addition, Bossel (1998) identified a set of common characteristics applicable to all kinds of systems, namely existence, effectiveness, freedom of action, security, adaptability and coexistence. To understand the potential of technological changes it is important to determine whether one, or more, of the Bossel criteria pose a definite obstacle to the introduction of a new system component, or if there is merely a temporary mismatch (the following set of indicators is suggested, see Table 4-7).

Table 4-7. Economic indicators of sustainable urban water management.

Stakeholders	Normative conditions	Indicators
Users and Organisation	The users demand affordable service and are able to pay it.	Proportion of household disposable income spent for water and sanitation (% of disposable income).
Users and Technical structure	The users' demands must be compatible with the technical capacity.	Time and money spent on operation by household or property owner (Euro per person, or time spent per year).
Users and Technical structure	If a system is changed, users must be willing to pay for it.	The users' willingness to pay for alternative technical structures.
Organisation and Users	An organisation must be able to recover costs for construction, operation and maintenance.	Collected fees in relation to costs for operation and maintenance, i.e. cost recovery (%). Funds generated internally in relation to investment costs, i.e. the rate of internal financing (%)[1].
Organisation and Technical structure	An organisation must use its resources in an efficient way.	Unit volume costs expresses as natural resources, labour and money (Annual cost per volume of water, or per person).
Technical structure	The costs of construction, operation and maintenance.	Use of natural resources, labour and man-made capital for construction, operation and maintenance (Annual cost per user).
Externalities	The negative impacts of the system should be minimised. The positive impacts of the system should be maximised.	Costs and benefits in other parts of society and the environment (CBA, MCA).

4.2.5 Scales

In working with indicators, both the geographic scale and the time scale play a significant role.

Things that matter on a local scale may be less important on the global level, and vice versa. Although water pollution is a problem on any scale, the negative impact may vary.

Point source emissions usually have a strong local impact, while diffuse loads from agriculture or from air transports have more widely spread effects that are difficult to counteract with remedial measures. Indicators of pollution do not always coincide with politically defined borders, such as cities, municipalities

---

[1] This requires that some of the income from water rates, for example, is not allocated to the operation of existing plants.

and counties, which have hitherto been responsible for decision- making in the water sector.

Through the EU Water Framework Directive (2000), a new administrative level is introduced: catchment districts. The district authorities should see to it that water resources are kept in good condition and utilised soundly. This implies that flows and loads from all sectors must be coordinated. Substance flows from agriculture, urban areas, industries and forestry will be assessed together, and economic analyses of water usage will be performed (Löwgren 2003). Remediation programmes are to be established as a part of management plans, to assure feasible and cost-effective improvements of water quality. The objective is to reach good ecological and chemical water status by 2015. This objective might be too optimistic in many areas, both because of nature's delayed response to pollution abatement measures, and also due to societal and institutional inertia.

Consumer attitudes and lifestyles are changing slowly. This fact draws attention to the need for a deep understanding of the processes of change, both in the eco-systems and in society. Catchment modelling of nutrient flows and a better understanding of how to use various incentives for pollution control are valuable complements to sustainability indicators (VASTRA 2005).

5

Assessment: sociocultural aspects

5.1. Henriette Söderberg and Mats Johansson
5.2. Helena Krantz and Jan-Olof Drangert

5.1 INSTITUTIONAL CAPACITY: THE KEY TO SUCCESSFUL IMPLEMENTATION

Sustainable urban water systems are more than a technical issue. They also represent a trans-disciplinary challenge and raise the question of institutional capacity for implementing sustainable systems, as well as the operation and maintenance of such systems. It is within the planning process that the prerequisites for successful implementation and maintenance are shaped; here, these are denoted institutional capacity. This chapter concentrates on criteria for institutional capacity: What are the critical factors related to institutional capacity? How can these criteria be used in practice? What kind of tools can support comparative assessments of alternatives where institutional capacity is concerned?

© IWA Publishing 2006. *Strategic Planning of Sustainable Urban Water Management* edited by Per-Arne Malmqvist, Gerald Heinicke, Erik Kärrman, Thor Axel Stenström and Gilbert Svensson. ISBN: 1843391058. Published by IWA Publishing, London, UK.

Criteria are used here to describe demands to be met. This approach is inspired by system theory and human ecology. Indicators are, using the same language, the degree of fulfilment of each criterion. For institutional capacity, we work only with criteria, not indicators. The way to measure degree of fulfilment can differ with respect to place and time. There are no quantitative thresholds for criteria related to institutional capacity, as there are for environmental aspects or for hygiene issues. It is therefore worth noting that the kinds of criteria presented here are not useful for benchmarking. Deciding whether institutional capacity is good enough is time and place specific, a task for the decision-makers while they are conducting their assessments.

Decision support tools have generated great interest among researchers and politicians as a strategy for handling complexity in decision-making. One of the purposes of these tools is to make decision-making more systematic when several alternatives are at hand. Decision-support tools can be more or less complex computer models, but they can also be graphical tools such as pictures or checklists. For institutional capacity, several kinds of graphical tools can be used to show differences between alternatives, or to generate a common understanding among the stakeholders. Examples of these are presented in the end of the chapter.

Although a helpful picture may look simple, it can nevertheless be very useful. The aim is simplicity without triviality. The management of complexity in the planning process is also described in Section 8.

5.1.1 The criteria

The criteria introduced here are not specific for the urban water sector and can be used as a checklist for the successful implementation of most planning processes. The waste management and mobility management sectors have been using similar criteria for some years. The criteria originate from three Urban Water model cities, and from the literature on organisational theory, planning theory, and theories about common pool resources, such as water and air, as well as from other studies of the urban water sector in Europe carried out during the past decade. Seven criteria have emerged that comprise the most critical demands to be met:

- The presence of policy entrepreneurs, *i.e.* initiators as well as implementers;
- The sphere of action, such as legislative and political support;
- A value coalition of shared world views, problems and goals among crucial actors;
- Access to resources such as knowledge and money;
- Explicit division of responsibilities and risks among actors involved;

- A defined arena for participation and conflict management; and
- Communication with users.

It is recommended that the seven criteria be considered thoroughly when planning for urban water systems. They can be divided into two types: criteria more important in the beginning of the planning process (the first four) and criteria more important during the planning and implementation process (the last three).

The presence of policy entrepreneurs: initiators and implementers

In strategic planning there is a constant need for what can be called *policy entrepreneurs* in the different phases of the process (Bossel 1997; Hjerpe and Krantz 2001; Blomqvist 2002; Ostrom 1990; McKean 1992; Rodenburg *et al.* 2001). It is seldom the same actor who sustains forward momentum during both the planning phase and the implementation phase. To make the process less vulnerable to policy entrepreneurs leaving the stage, one option is to build a dedicated working group; another is to create awareness of the importance of actors ready to take on the responsibility. Do not underestimate the importance of building a platform for the process and planning for bridging between phases! Policy entrepreneurs can sometimes be found outside the formal organisation, or within the organisation but without having formal power. If managed properly, this is not a problem. On the contrary, make use of the energy the policy entrepreneurs can offer!

Sphere of action: Legislative and political support

Implementation requires a sphere of action, which can be used in different ways. If need be, it can be generated, but without it the process is lost (Hardi and Barg 1997; Söderberg and Kain 2006; Blomqvist 2002; Rodenburg *et al.* 2001). The state or a financial institution can provide a sphere of action by means of financial incentives or by legal restrictions. This kind of sphere can also be generated by allocation of resources. Here the policy entrepreneur is important. Legislation can both establish a sphere of action and delimit it, as more or less defined by legal restrictions and support from politicians or company boards. *More or less* refers to the possibility of interpreting the given legislation.

Overall, it can be concluded that the legal prerequisites for realizing the system structures studied in the Urban Water program are fairly straightforward (Olofsson, A. 2004). However, the final design of the systems must meet the specific technical requirements that are regulated by legislation. In particular, the treatment processes must meet requirements for effectiveness and adaptability to future demands. In the source-separating wastewater systems, the idea of using untreated human excreta, *e.g.* urine, in agriculture has few legal hindrances in Sweden, but remains problematic for other reasons.

The value coalition between crucial actors: A shared world view, problems and goals

A value coalition among key actors is very useful for large projects such as urban water systems (Ostrom 1990; McKean 1992; Söderberg and Åberg 2002; Rodenburg *et al.* 2001). This coalition can be shared worldviews, goals in common, a shared problem definition or a mutual interest in incentives. An established value coalition is a strong driving force that contributes to continuing according to plans. However, it is equally important in working together to make it clear to everyone involved what views you do not share. Differences can be managed, but more easily so if there is an awareness of their character. A warning must also be raised here: do not consume a lot of resources on establishing value coalitions among actors who are unenthusiastic. Investigate what key actors have in common and what disagreements exist, and then move on. Important is that the issue has been reflected upon.

Access to resources such as knowledge and money

Allocation of resources is crucial, not just financial resources but also access to knowledge, both scientific and practical experience from demonstration projects or related projects (Hjerpe and Löwgren 2002; Rodenburg *et al.* 2001). New solutions always involve anxiety among actors who do not want to be first to make all of the mistakes. The recommendation here is to communicate with other actors who have related experience, and to include second opinions in the briefing material during the planning process. It has been found valuable to have key actors with a positive scepticism attached to the planning of a project. They can point out the weak or questionable aspects of alternatives in an early phase.

The business administration view upon sustainability, as applied at Lund University, defines an organisation as sustainable when it has the ability to allocate and maintain resources of three kinds: financial resources, competence and natural resources (Thomasson *et al.* 2005).

Explicit division of responsibilities and risks among actors

In the Urban Water approach, the definition of a system does not only focus on the organisation that has the formal responsibility; it includes the users as well. In an analysis of the institutional network, it is important to include stakeholders of various kinds. Alternative urban water systems can have very dissimilar patterns of responsibilities, with a large international multi-utility company as one extreme, and a group of neighbours managing their local system as the other.

The question of whether sustainable urban water systems are more easily obtained by multi-utility companies with large resources, or by local organisations of neighbours managing their water and wastewater systems together, does not have a general answer. Instead it is, again, time and place specific conditions that determine what to do. Important in the beginning of a planning process is to be open-minded towards the options at hand, as they are often many more than expected.

A critical prerequisite for management as well as for operation and maintenance is a clear division of responsibilities, including financial responsibility (Söderberg and Kain 2006; Hjerpe and Krantz 2001). The latter has a close connection to risks, financial and others. When implementing infrastructural projects, there are often financial hindrances during the process; other hindrances can be related to who should take responsibility for operation and maintenance. All of these cannot be foreseen, but they can be addressed more systematically if awareness of their possible occurrence is built into the planning process in an early stage.

Here, network analysis can be useful, and an array of graphical tools can be used to make the process clear to everybody involved, especially what specific responsibility each actor has (see Figures 5-1 to 5-3 below).

Arena for participation and conflict management

A planning process that gives users and stakeholders a chance to participate provides prerequisites for system that functions well (Healey 1997; Ostrom 1990; McKean 1992; Hjerpe and Krantz 2001; Rodenburg et al. 2001). While this does not mean that everybody involved has to participate, offering the option to do so means there are also arenas for conflict management to smooth the process. Participation and communication do not guarantee trust; however they can decrease the risk of distrust. This is closely related to the criterion of a value coalition between users, since an arena can be the place for discussions of the common views of and goals for a project.

Communication with users

Well-functioning urban water systems were shown to have high quality interaction between the users and the technical system structure, as well as interaction between the users and the responsible organisation; a strategy for communication with the users of the system is therefore important (Söderberg and Åberg 2002; Rodenburg et al. 2001). The users are central to the success of the systems, and technical design communicates, as well as written or oral information. As with most of the other criteria presented here, main importance is to raise the awareness of the importance in a well-planned communication process with the users.

The first four criteria are the most critical in the very beginning of a process. A policy entrepreneur, whose work is initiated with allocation of resources, can generate a sphere of action and a value coalition. In contrast, a value coalition can also allocate resources and seek out a policy entrepreneur. Division of responsibility and risk, which needs to be worked out early in the process, will require closer management later during the process.

How to use the criteria

The criteria are to be used to guide the gathering of information for the briefing material handed out prior to meetings in the early stage of a planning process. They are also needed for comparative assessments of alternatives. As mentioned before, there are no thresholds connected with these criteria; it is up to the decision-makers to decide what is good enough. There are often goals or ambitions that can be used as reference values.

When a criterion is not fulfilled, there is no need to completely disqualify the alternative studied. Another choice would be to concentrate on the factors that cause the low institutional capacity and to turn them into opportunities. This may improve institutional capacity, if the means are available to the actors. The criteria can also be used during the process as a checklist for factors to consider and discuss, for example when a problem arises. The criteria could be adopted as guiding principles for designing a high quality policy process that comprises all of the phases of planning, decision-making and implementation.

5.1.2 Experience in using the criteria

The seven criteria are, as for the other Urban Water criteria, to be viewed as guidelines that can be adjusted to a specific case. It is essential that the criteria chosen be perceived as relevant to and understandable in the situation to which they are applied. Within the Urban Water programme, we have used these criteria in two of our model cities, the small town (Surahammar) and Uppsala. The criteria were also used for the Sandviken and Södertälje project presented in section 8.

A consultant often has to find a technical system structure suitable for a specific situation in which water and wastewater are perceived as problems. The project usually results in one or a few technical solutions; these are the ones the consultant knows best beforehand, although they may not fit the local circumstances. To deal with the soft aspects of the process, the consultant or external actor needs support and instruction from the project owner. There is a need for support tools and methods to facilitate a structured discussion of the organisational and institutional aspects. Support tools are needed:

- in the early stage of the planning process when the problems are formulated and alternatives shaped,
- in decision-making situations where the comparative assessments between alternatives are made, and
- in parts of the process where stakeholders are involved.

The case of a small town: Surahammar

The Urban Water model city of Surahammar is described in detail in Section 9.2. For institutional capacity, three of the seven criteria were selected as the most relevant. They were division of responsibility and risk, assessment of the sphere of action and allocation of resources. A starting point for selecting criteria was that they should yield different results with respect to the system structures studied.

The main conclusion from Surahammar applies to the division of responsibility and risk. The network analysis used to guide the assessments was perceived as too complicated. There was not any reference point for the criteria; just to make the division of responsibility clear to everyone was not enough according to the local group of actors. A complication here was that the municipal service company retained the main responsibility, despite the change of system structure; however, the network analysis indicated that the blackwater system affected more actors than the present conventional system. The local group in Surahammar made the assessment of the sphere of action from the legal and political perspective without hesitation. In contrast, the sphere of action was discussed much more for the integrated planning in Uppsala.

Integrated planning in Uppsala

The integrated planning in the model city of Uppsala is described in detail in Section 9.5. In Uppsala, the local group of actors set priorities among the criteria for institutional capacity: they chose value coalition, sphere of action and policy entrepreneurs. The sphere of action was redefined as legal aspects and the formal planning procedure. The local group consisted of the heads of municipal departments affected by urban water systems; they considered it speculative to assess the issue of political support.

In Uppsala there was also a discussion about the circumstance that the criteria for environmental aspects, as well as for hygiene, apply to systems already implemented, while the criteria for institutional capacity deal mainly with the implementation phase, not with operation and maintenance of the systems. This is of course a complication, but it is not a valid argument for excluding these kinds of criteria, which often is raised as a suggestion. Implementation is a major challenge for sustainable urban water systems, and the implementation starts with the present situation whether the project continues for a year or for a decade.

Kullön, Vaxholm municipality

The seven criteria presented were used in a regional project outside Stockholm in 2004 (Richert-Stintzing *et al.* 2005). The institutional analysis was made when the first building phase was already completed, and people were living in the houses and using their urine diverting toilets. With the second building phase about to start, the operation and maintenance and implementation phases were combined.

In this situation, the criteria were helpful in identifying weaknesses in the existing local organisation and in the division of responsibility. If the central actors had made an analysis based on these criteria earlier in the process, several of the problems they had to manage later on would not have arisen. This was also important, because many of the decision-makers and key actors were not the same as those in the planning phase; the new people were carrying on the work with unclear division of responsibilities and without a shared problem definition or common goals.

When system structures involve actors such as real estate owners, farmers or entrepreneurs other than the local municipal offices, a structured analysis of the actors involved and their responsibility at different stages of the process is needed. Even if all the criteria presented here do not have to be fulfilled in order to obtain sustainable urban water systems, one may conclude that they make the road towards sustainable development smoother if considered and managed.

5.1.3 Graphical support tools

Several graphical tools have been developed to illustrate institutional capacity and to support comparative assessment between alternatives with respect to the seven criteria already given. They also help the group to stick to the important points during discussions. For the model cities mentioned above, graphical tools were used, which offered experience and revealed the need for further refinement. The figures here are offered as examples and ideas of how to work with graphical tools in a planning process; this is why there are no detailed instructions with them. The details are not as important here as they were in a specific planning process.

In the planning situation, the way that actors are related to the process can be divided into categories, such as members of the public, private, and civil society. The advantage of this kind of mapping is that one gets a better picture of the complexities and potential hindrances and opportunities. A disadvantage is the lack of information about the interrelationships, *i.e.* how the actors relate to the situation.

To illustrate the interrelationship between actors, a network can be sketched, illustrating which actor communicates or interacts with whom. To make a network of actors is self-instructive: take a sheet of paper and start! This kind of picture is useful to identify the links between people, which are sometimes self-evident, and sometimes need more structured investigation to be identified. It is useful to make the responsibility that actors have in the policy process explicit for everybody involved. Actors can be grouped according to categories of responsibilities.

The types of responsibility for the planning process, such as the financing, initiation, operation and maintenance, as well as regulation are shown in Figure 5-1. This kind of diagram was used in both Surahammar and Uppsala (Sections 9.2 and 9.5). Figure 5-1 illustrates the organisational system structure for blackwater in Surahammar. The main purpose of using this type of diagram is to show an organisational system structure, a picture of the actors involved, the type of category they belong to, and what kind of responsibility they have. Each actor can have several responsibilities, which may change over time during the process.

The disadvantage of diagrams like Figure 5-1 is their static nature. Another type of diagram could follow crucial actors through the planning process. This was done for a project at Kullön, Vaxholm (Figure 5-2). The four phases of the process are distinguished: the planning, the decision-making, the implementation, and the operation and maintenance. Figure 5-2 can also be used as a sorting tool, by placing actors into the phases that involve them. Categories of actors are listed in the figure, and their importance can be indicated by size of font. Again, this is not obligatory; it is simply an idea for managing the division of responsibilities and to make the situation clear for everybody.

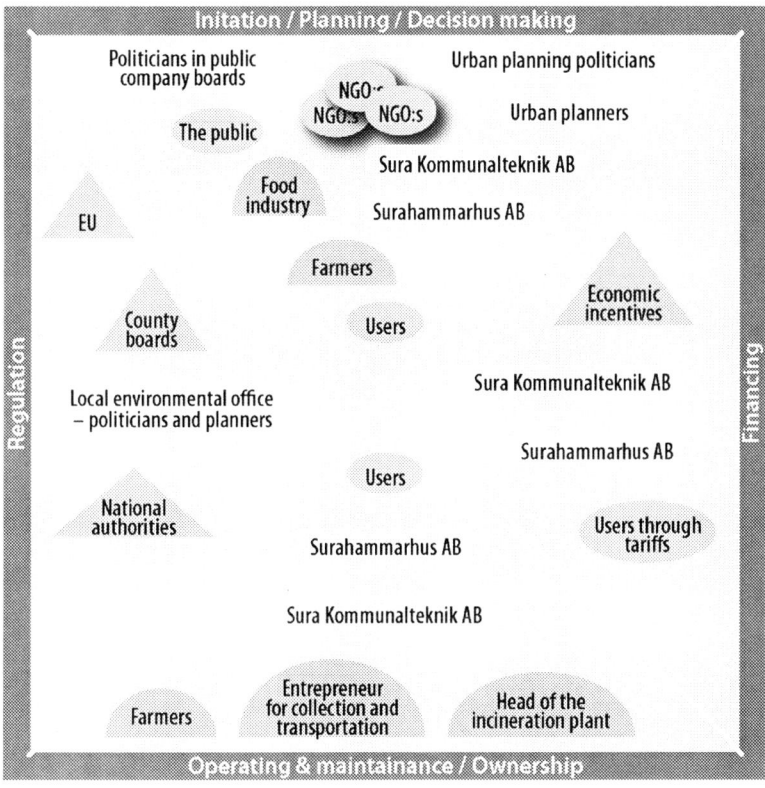

Figure 5-1. Example of the division of actors according to type of responsibility. An actor can appear in all four fields. The categories of actors can be indicated by shape and colour. Abbreviations in Figure 1: SuraKommunalteknik AB is the municipal multi-utility company responsible for water, wastewater and waste treatment. Surahammarhus AB is the municipal housing company. A county board is a central authority at the regional level that coordinates the various interests of the county.

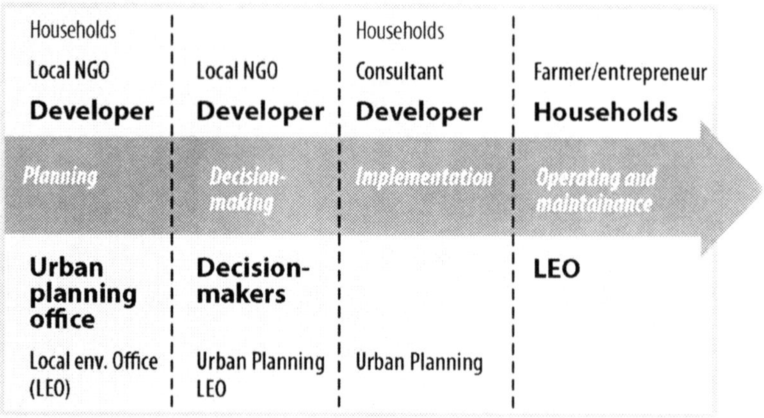

Households		Households	
Local NGO	Local NGO	Consultant	Farmer/entrepreneur
Developer	**Developer**	**Developer**	**Households**
Planning	*Decision-making*	*Implementation*	*Operating and maintainance*
Urban planning office	**Decision-makers**		**LEO**
Local env. Office (LEO)	Urban Planning LEO	Urban Planning	

Figure 5-2. Example of actor mapping through a process over time.

One can also relate the actors to specific parts of the technical system structure (Figure 5-3). This approach can be useful when choosing what alternative to implement. Then, a plan for the implementation phase can be supported with a clear picture of actors to be involved.

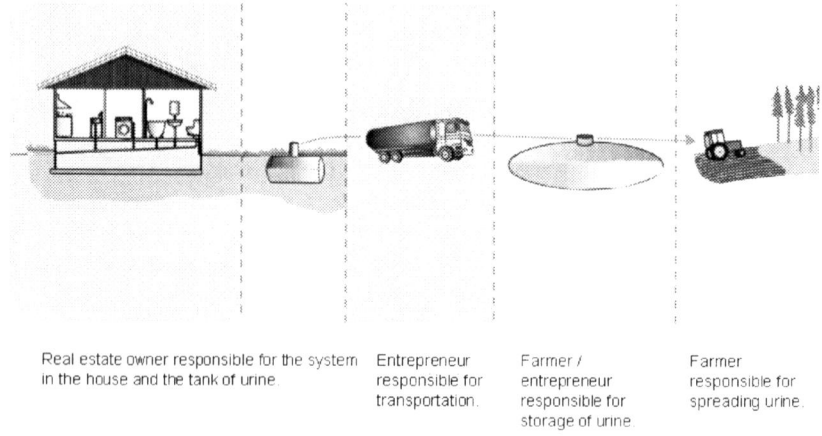

Real estate owner responsible for the system in the house and the tank of urine.   Entrepreneur responsible for transportation.   Farmer / entrepreneur responsible for storage of urine.   Farmer responsible for spreading urine.

Figure 5-3. Mapping actors in relation to the technical system structure. Here an example with handling of urine in a urine sorting system. © Verna Ecology.

Table 5-1. Graphical tools and their application.

Part of the process	Graphical tool	Critical questions
1. Early phase of the planning process	Actor mapping: Network of actors	What status does the project have? Does the board of the water utility or the local politicians sanction it? What actors shall be involved? What is their role?
		Are there external processes and actors that can or will influence or hamper the project?
2. Shaping the alternatives	Organisational system structure: division of responsibilities (Figures 1, 2)	Are all crucial actors considered? What aspects are most relevant to discuss with them: system design, economy or O&M?
3. Decision-making	Possible to chose between alternatives according to the given conditions. Diagrams can be very important as a starting point for decision-making.	What diagrams used earlier in the process are appropriate or can be re-drawn? Are all institutional aspects of importance communicated and the consequences of each alternative correctly understood by the decision-makers?
4. Implementation phase	Planning for implementation (Figure 3)	What actors have the main responsibility in the course of the process?
	Division of responsibilities over time (Figure 2)	Do the actors responsible for future operation and maintenance have the competence for the alternative chosen?
	Go back to the original situation in the early planning phase (Figure 1).	Have actors, their roles, or external aspects changed since the start of the project?
5. Operation phase	Diagrams 1, 2 and 3 can show the final organisation and responsibilities for operation and maintenance.	What kind of back-up exists? What kind of arena for problem solving and discussions is there? Who is responsible for keeping contact with all relevant actors?

Table 5-1 can be extended to contain also strategic methods that may be appropriate to the project, as well as strategic institutional milestones or decisions that are needed before moving on to the next phase.

5.1.4   Recommendations

When planning for sustainable urban water systems, or any other infrastructure system, there is a need to integrate the issue of institutional capacity early in the process. Institutional capacity should be given the same priority as other kinds of knowledge. Relevant information about institutional aspects should be included in the briefing material for meetings. Include it also when comparing alternatives, but treat each kind of information according to its own frame of logic: they should be dealt with separately. Use graphical tools to support communication.

There is little relation between the complexity of models or illustrations and the value of the information. This chapter has introduced the reader to helpful criteria for institutional capacity and provided ideas on how to use graphical tools in a planning process.

5.2   HOUSEHOLD PERSPECTIVES IN MANAGING SUSTAINABLE CITIES

5.2.1   Introduction: the road to where we are now

Utilities and households have had to manage water and sanitation ever since the piped water supply and sewerage were installed in Sweden in the late 19th century. Starting then, all water-related household activities were done in the home, which meant a revolution for household work: no more fetching of water at the well or river, no washing of laundry in river water, and no disposal of buckets of used water. Here we can find the origin of the present norm that water and sanitation should be comfortable and involve little work.

The welfare state supported the emerging service democracy. Until the 1970s, the relationship between households and utilities was one of utilities providing services while residents only had to pay. The households were anonymous subscribers with no responsibility for what was discharged to the sewerage. The bills were based only on the quantity of water used, as if there were no cost attached to the treatment of wastewater or environmental degradation. This was also reflected in the reaction that only utilities needed to take action when surface waters were heavily polluted with wastewater (in the 1940s and 1950s), while residents had no role to play.

In Sweden, this state of affairs changed rapidly when the new water act came into effect in 1971. The act stipulated that the utilities had to cover their costs, and no subsidy from ordinary taxes was permitted. This occurred in a period when the cost for wastewater treatment rose sharply. The utilities started to view subscribers as customers in an effort to make them accept higher prices for water-related services. This shift required that the customers should be informed about what services they were provided and why they had to pay more. The bills now detailed the cost for water supply and for wastewater treatment.

The 'customer period' was successful in that residents soon paid for actual costs, except for the subsidies provided by the government to build treatment plants and sewers for isolated villages at a distance from the city centre (Drangert and Löwgren 2005).

During the 20th century, new chemical products entered the market at an alarming rate, and after use a lot of them ended up in the wastewater treatment plant. It was increasingly realized in the 70s and 80s that the wastewater treatment plants could not cope with the challenges of treating this wide range of chemicals and of recycling the nutrient-rich sludge safely to farmland. The magnitude of the task is indicated by the fact that the number of chemists in utilities is but a very small fraction of the number employed by producers of chemical products such as pharmaceutical and detergent manufacturers (Palmquist 2004b). This is also a time when the environmental legislation is becoming tougher and utilities fail to fulfil the new requirements (EU Reach Programme 2005).

For the first time in a hundred years, the utilities see the need to involve residents to solve a problem. The main role anticipated for residents is to reduce the load of hazardous substances in wastewater by applying source-control measures to discharges from the homes. In the late 1990s, utilities began to inform households about what can be flushed down the toilet and sink, and what products are not allowed to be flushed away. This shift in focus from what comes into the home to what goes out of the home makes residents true partners in the management of sanitation.

At this stage, the Urban Water research programme looked into what residents do and why. The following presentation of the results has a prospective approach rather than a retrospective one.

5.2.2 Households as partners

The household as a partner has a defined role, such as sorting solid waste and bringing harmful leftover medicines to the pharmacy, reducing the volume of hot water use, and installing mixer taps. Still, residents say they are uncertain about their role as a partner. Their knowledge of how the large and invisible water- and energy systems affect the environment is limited. The professionals at utilities, on the other hand, are not yet eager to shift some of the responsibility to the residents or to involve residents in the development of new technology. They still employ information as the main strategy to influence household routines (Rydhagen 2003; Drangert *et al.* 2005). This shows that the shift of households from customers to partners requires a negotiated agreement between the partners at appropriate levels.

However, change is on its way and alongside utilities' information efforts we also witness housing companies taking on a more prominent role by increasing the use of economic incentives such as volumetric billing of hot and cold water, and technical installations that help to enhance households' ecological performance without requiring more work or changing routines. Such measures are not controversial, since they are in line with residents' preferences. In general, residents are in favour of billing according to use, both for the prospect of economic gain and out of fairness (Axelsson *et al.* 2001; Drangert and Krantz 2002; Krantz 2005); although residents find resource saving important, it should be built-in and function without infringing on everyday life (Åberg 2000 and 2004; Drangert *et al.* 2005).

The focus is slowly shifting from input of materials to discharges of used materials. Due to raised requirements for sustainability, there is a continual change in our perception of material flows. The word waste is used at times when communities fail to arrange effective systems for handling residues from production and use. There are four main groups of physical products entering households, being consumed there, and most of which are discharged as wastewater: energy, water, food and consumer goods (Figure 5-4).

Earlier on, mostly sewer workers and refuse collectors knew the output from homes, whereas residents tended to view the incoming products and consumption patterns as the relevant issues. Environmental awareness is partly about a change in focus from input to output issues of our cities' material flows, while accepting that the two are intrinsically intertwined (Drangert 2004).

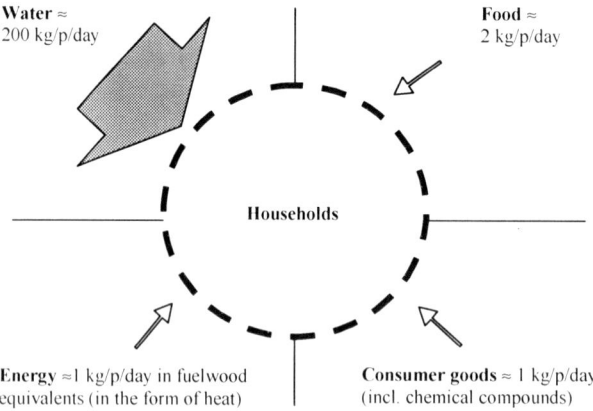

Figure 5-4. Input of various materials to the household in kg per person and day. 1 kg of fuelwood equivalent = 13.8 MJ (adapted from Drangert 2004).

5.2.3 Built-in measures to improve sustainability

Professionals and residents have a common wish for equipment and installations that are easy to use correctly, and also that these should make abuse of the system less likely. For instance, previously we had to turn on one cold and one hot water tap, and adjust the combination, to get the desired temperature. With the new mixer tap we can save water by easily turning it on and off while brushing teeth, shaving or showering, since the water temperature remains the same. However, the resting position of the mixer tap handle is often in the hot water position, which means that more hot water is being used than for a resting position that requires the user to turn the handle to get hot water.

Another example is the meter for reading the household use of hot and cold water. The purpose is to lower the quantity of water used by keeping the household informed. However, our studies point to the difficulty in reading and understanding the meter and the utility invoice. Consequently, few households use the information (Drangert *et al.* 2005; Krantz 2005). These are some examples of poor technical design that inhibit good habits.

The sanitation arrangement is easy to use, but also to abuse by throwing any harmful liquid or solid waste into the washbasin, sink, and flush toilet. There are no technical devices to make such practices more difficult than to bring the product to the proper collection place, except for the dimension of the sewer pipe and the strainer in the sink. Having a waste bin in the bathroom reduces the amount of solid waste going into the toilet. Also, having enough space in the cupboard under the kitchen sink to collect organic and other solid waste makes it easy to avoid throwing unwanted items into the sink (Figure 5-5).

The installation of a kitchen waste grinder removes the task of bringing the organic waste to the recycling room, and householders regard it as a convenient solution for handling the kitchen waste (Rydhagen 2003; Åberg 2004). On the one hand, it will ideally collect all organic waste and transport it to the wastewater treatment plant where it helps to digest the wastewater by adding carbon. The grinder gives some feedback on disposal behaviour, since it cannot grind plastic, bottle caps and the like (Åberg 2004). On the other hand, the presence of a grinder makes the user believe that he or she may throw more solids into the sewer than what was thought before. The likelihood of putting harmful material into the grinder either by accident or on purpose is assumed to be increasing.

An important investigation concerns the extent to which built-in arrangements will improve ecological sustainability. The average daily household water usage in Sweden is some 190 litres per capita, but in new houses it is 150 litres.

Studies of the newly built Stockholm suburb, Hammarby Sjöstad, show that built-in arrangements have reduced cold-water use by 50% when compared with a standard house (Drangert *et al.* 2005). A comparison of a collective tenant owners' association with ecological aspirations, dry toilets and shared water bills, and a conventional housing area with flush toilets and volumetric billing, showed that households in the ecological collective tenant owners' association generally used less cold water but more hot water. The lower cold water use was explained by the dry toilets, while the higher hot water use was caused by a combination of a lower hot water temperature (also contributing to the lower cold water use), fewer water-saving showerheads and taps, and more families with children (Krantz 2005). Built-in arrangements do save both hot and cold water.

Figure 5-5. Waste collection boxes for organic waste and other waste (left); and a waste bin for solid waste in the bathroom (right).

No study has been found that identifies what harmful products go into the wastewater from households and entering the wastewater treatment plant. Investigations have focused on the incoming and outgoing water from the treatment plant. This gap in knowledge has to be acted upon in the process of defining the partner role for households. Some more background information is expected once the EU legislation on chemicals (REACH) is introduced. In the meantime there are additional measures and arrangements that can be carried out locally to save on resources and to improve wastewater quality and sustainability (Drangert *et al.* 2005).

5.2.4 The time has come for adjustments of household routines

Built-in arrangements have been successful up to now in saving resources, but they do not seem to impact hygienic norms or preferences for comfort. Water and chemical products are used in an unchanged manner to provide services such as clean clothes and an odourless (or perfumed) body perceived as necessary to make the right impression (Krantz 2005). If we want to reach further in saving water and increasing the quality of discharged wastewater, the residents have to be allocated a clear role. The challenge is to find out what circumstances will lead householders to take on new responsibilities and adjust their routines to improve sustainability.

A household is not a well-defined unit, and may consist of a single person or a large number of people of varying ages. Also, householders may reside in a flat or a detached house. Since the residents in privately owned detached houses have more responsibilities for the infrastructure than tenants, it could be informative to study them. The Urban Water programme, however, has exclusively studied households in flats in various types of housing areas (old city centre, a suburban area with tenants' houses, a newly constructed part or suburb of a city, small town, and a collective tenant owners' association). People move, however, and a large portion of the residents in Hammarby Sjöstad (new town area), for example, previously resided in detached houses in the outskirts of the city. They are expected to be aware of and careful with resources, such as water and energy, since they have paid according to usage. Tenants in flats with shared bills for water appear to be less familiar with and sensitive to the relationship between usage and cost. The expected situation is depicted as shown in Figure 5-6.

The new residents in Hammarby Sjöstad encounter resource-efficient technical arrangements and their usage decreases, but they may react differently on the charge for the services. Since the cost for energy and water is shared among all, residents moving in from a detached house may increase their water and energy use (info 1 in Figure 5-6). For those moving from houses, it would be more difficult to lower the already low amounts previously used than for those moving in from another flat. As time passes the usage may increase, and there is a need to repeat the message of saving (info 2 in Figure 5-6). There is a negative attitude among all residents towards wasting water, while saving depends on lifestyle requirements (Drangert and Krantz 2002). Any measure to change household routines has to take such variations into account.

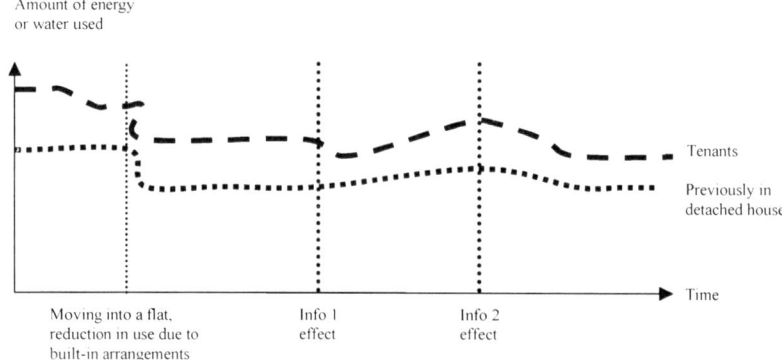

Figure 5-6. Presumed impact of built-in arrangements and information activities on household usage of water or energy.

Households do not consume water, products and energy; rather, they use them to achieve their desired lifestyle. For example, there is a rather recent norm in society that shirts and blouses should not be worn more than one day before being washed. Frequent washing of clothes is a way to fulfil expectations. It is therefore not negotiable to wear clothes for more days in order to save water or to prevent deterioration of wastewater quality. The two ideas are simply not compatible. Variations in water-use routines seem to be largely socio-culturally defined. There is a lower limit for water use, regarded as necessary for maintaining sufficient standards of cleanliness and comfort, irrespective of ecological and economic incentives (Krantz 2005).

Despite the cultural imperative, the physical representation of a rest-product affects or moulds peoples' perceptions of their role in the resource flows. The more visible a rest-product is, the more the household acknowledges its responsibility. Energy does not leave any trace in the home after use, while wastewater is seen briefly before it disappears into the sewer. However, products bought and brought home from the grocery shop are seen also after use. Residents do not expect anyone but themselves to sort their solid waste, while they expect the wastewater treatment plant to be capable of treating all their wastewater irrespective of content (Axelsson et al. 2001; Drangert and Krantz 2002; Drangert et al. 2005). Similarly, residents in the collective tenant owners' association did not expect anyone else to manage the excreta from their urine-diverting dry toilets. Perhaps more important, this individual responsibility for operation and the knowledge that faecal matter and urine is returned to nature resulted in environmentally friendly routines where only biodegradable waste was thrown into the toilet and environmentally friendly detergents were used for cleaning. No information campaign for the proper use of flush toilets has been so successful (Krantz 2005). Åberg (2004) also points to the importance of motivation when residents are able to make positive associations between their knowledge and what other residents are doing.

Increased technical transparency or visibility of responsibility seems to support environmentally friendly routines. For instance, individual metering and billing of hot and cold water make visible a cost, previously hidden in the rent. A resident can now influence the bill by adjusting his or her own use. Volumetric billing was introduced as part of a large-scale refurbishment of a low-income area (Ringdansen). Obviously, low-income households put under the impact of volumetric billing do consider moving to another area with collective billing. In Ringdansen about 25% of the households returning to their refurbished flat later on chose to move (Hjerpe 2005). Among those staying, hot water use was reduced by 10 to 20% in one of the blocks and by 20 to 40% in another. Corresponding figures for cold water were 15 to 25% and 15 to 35%. Eventually the two blocks showed similar average usage per capita and day. The difference between the blocks is partly explained by that a higher proportion of high-consumers moved out, soon after the refurbishment, from the block showing lower reductions than from the block showing higher reductions. Water is generally regarded a cheap commodity, and for a saving effort to be worthwhile, it should be possible to save a significant amount of money (Drangert *et al.* 2005). This often applies to high consuming low-income households. From this it follows that they tend to make greater efforts to lower their usage (Hjerpe 2005; Krantz 2005).

Changes in water and sanitation arrangements leading to increased system transparency seem to induce routine changes. The more extensive the change in arrangements is, technical or physical, organisational or economical, the more radical are the routine changes that follow - and more so if measures are combined. However, the improvement of ecological sustainability is still conditioned by what is socio-culturally accepted. Often the more extensive changes in arrangements, for example dry toilets, are more critical in this regard (Krantz 2005).

Residents do not just passively accept water and sanitation arrangements but also influence them to a certain extent to make them fit with desirable routines (Krantz 2005). Therefore, better and more communication between residents and technical designers, as well as greater resident participation in decisions on system solutions, could contribute to resident acceptance (Rydhagen 2003). While this is valid for places where people are just moving in, it does not take into account second-hand contract holders or residents moving in later on (Drangert *et al.* 2005). Housing companies, real estate agents, water companies etc. have a long-term role to play in this regard. Concerned stakeholders and actors must be provided the opportunity to participate in the planning process, and there should be arenas for communication between different kinds of actors (Storbjörk and Söderberg 2003). However, consensus in such participatory processes is hardly possible, and the goal should ideally be to reach a decision that is tolerable for all parties (Söderberg and Kärrman 2003).

5.2.5 Information activities

As residents become partners, a new perspective is emerging. Seeing to it that the residents are informed and aware becomes as important for utilities as maintaining the physical installations. Utilities have begun to inform households about what can be flushed down the toilet and sink, and what products are not allowed to be flushed away. Still, households are uncertain about their role as partner and they request more information on do's and don'ts. In our studies residents have suggested how the information could be designed to be effective.

Effective information according to residents' suggestions

- the role of the individual should be the focus of the information, as should the importance of the contribution of him or her;
- messages should be give in a positive form and not be top-down and demanding;
- information should be combined with financial incentives in order to increase the level of motivation and engagement of the individual;
- information should contain practical examples of measures that can be taken;
- information should be factual and provide analysis concerning the consequences, since it is difficult for the individual to assess all effects;
- information should be locally based;
- information should give feed-back on residents; efforts;
- information should be repeated regularly.

Box 5-1. Residents' suggestions for effective information. Adapted from Axelsson *et al.* (2001), Drangert and Krantz (2002), Åberg (2004) and Drangert *et al.* (2005).

The demand for more action-oriented information illustrates that residents are prepared to take on more responsibility if they have more information. Their demand could also be interpreted as a way for the informants to avoid criticism that they are not doing enough, since many are already making conscious efforts (Axelsson *et al.* 2001; Drangert and Krantz 2002). Information is needed to maintain residents' knowledge and motivation, and to build trust for the system. Printed information may be economically attractive compared to direct contacts, but many do not read it, and some messages are misunderstood. A huge challenge for all information providers is that the need for it varies between, as well as within, households[1] (Åberg 2004). The information should preferably be adapted to the perceived need and existing knowledge, but this is very difficult and costly given the heterogeneity of households. Therefore information and messages are often uniform and run the risk of not being sensitive enough to individual reasons for his or her actions (Drangert et al. 2005). Moreover, even if informing people is well done in terms of increasing the general knowledge about an issue, there is no direct relation between awareness of environmental problems and taken actions to reduce them (Åberg 2000; Krantz 2005).

There is scope for developing information packages that combine various measures and address the suggestions made by residents given in the box above. A comprehensive list of methods to inform is found in Drangert et al. (2005), which covers use and discharges of water, sanitation, solid waste and energy in households.

5.2.6 Recommendations for planning improved sustainability

Partnership relations require new utility strategies that put as much emphasis on communication with residents and housing companies as on the technical arrangement of water and sanitation. Residents must be part of the reuse and recycling management, and the strategy should include a greater physical and economic visibility of the use and discharge of energy, water and other products. The most powerful driving force for change is probably the all-invasive chemical society we are experiencing with approximately 30,000 different compounds used regularly in our homes.

As residents are prepared to pay for resource-saving installations, housing companies and other stakeholders should explore this possibility fully. Furthermore, most residents endorse the norm that it is not good behaviour to waste water and other resources. Since the issue of saving resources is more complicated, there is little consensus. Much of the resource use is for purposes such as cleanliness and status. Thus, changes in routines have to be negotiated with existing (but changing) cultural norms.

---

[1] Some participants in our studies have made decisive routine changes. For example, residents state that it was not until they started to write a time-diary about their household routines that they began to sort their kitchen waste, or use hot water more carefully, or contemplate what products they disposed of in the washbasin, sinks and toilets. Consequently, the methods we use as researchers to visualise household routines may increase people's awareness of their own routines. A researcher's presence and interest in these routines also highlight the relevance of communicating with residents by house calls and door-to-door approaches.

The heterogeneity of households as to age, number of persons per home, sex, ethnicity, housing type, knowledge and lifestyle preferences results in differing reactions among residents to taking action. For example, some already save water out of environmental concern and are only slightly affected by volumetric billing, if at all, while others are sensitive to economic incentives. All households cannot become partners by means of one single measure, however a variety of arrangements working in parallel could succeed. Households choose from these different arrangements by striking a balance between cultural norms and their willingness to engage and spend time to operate them.

Support to existing readiness to do something is likely to be more successful than to make uniform demands, which may be counteractive. One example is to introduce energy usage quotas. An individual household uses energy for heating the flat, heating hot water, light, transports, etc. If there is a requirement to lower the use of energy by half, the household could reach this by saving an equal proportion of all uses or let their preferences influence saving more on some and less on others. In this way, residents may be knowledgeable and responsible partners.

What it all comes down to is how we choose to live our everyday lives. Residents' activities and perception of the function of the home are constantly changing, and the shifts of the two do not always appear rational. Today, we prepare less food in our kitchens, while the standard of the kitchen is higher than ever. The size of the bathroom is expanding despite that washing is done elsewhere and the bathroom is increasingly used for relaxation. These rooms are also status symbols. The proposed diversity in technical arrangements could affect socio-cultural perceptions and break up standards of normality, since several kinds of arrangements could be perceived as normal. Diversity is an important ingredient when creating ecologically and socially sustainable water and sanitation systems and arrangements for the future.

6

Assessment: technical function

Gilbert Svensson

The urban water infrastructure includes drinking water treatment, the water distribution system, the wastewater collection system, wastewater treatment, and the stormwater collection system. The purpose of the water treatment together with the water distribution system is to supply the users with the amount of water demanded and to supply this water with adequate pressure under various loading conditions. The purpose of the wastewater collection system together with the wastewater treatment is to serve the users with a system that is capable of taking care of sanitary waste water, some drainage water and in many cases also stormwater under various loading conditions. The goal of the stormwater collection system is to serve the users with a system that is capable of taking care of stormwater and some drainage water under various loading conditions.

Much emphasis has been laid on the environmental performance and sustainability of water and wastewater systems; the most important issue, however, for the organisations that own and operate them, as well as for their customers, is that the systems work reliably. The issue of technical function has been systemised in EU framework research programmes, and applied also within some of the Urban Water projects.

© IWA Publishing 2006. *Strategic Planning of Sustainable Urban Water Management* edited by Per-Arne Malmqvist, Gerald Heinicke, Erik Kärrman, Thor Axel Stenström and Gilbert Svensson. ISBN: 1843391058. Published by IWA Publishing, London, UK.

6.1 THE DRINKING WATER SYSTEM

The central element of the drinking water system is to extract raw water from a source, treat the water to meet all relevant standards, and to distribute the water to the users without compromising its quality on the way. In Figure 6-1, the main components of a centralised drinking water system are shown; for the water distribution system, there are also some sub-components. There is normally a linear flow from the source to the tap. In regions with water scarcity, reuse schemes may introduce additional loops for water of lower quality for limited applications.

From both investment and operational points of view, the distribution system is the most costly component. To be able to meet drinking water standards at the consumers' tap, a water utility needs to give priority to the water quality during distribution.

The *Care-W*, a European Fifth Framework Research Programme, has defined a framework for the operation, maintenance and rehabilitation of water distribution systems based on performance indicators (Alegre *et al.* 2000; Alegre and Baptista 2003). These indicators, which reflect the objectives of the drinking water system, are divided into operational indicators, financial ones, the quality of service, water resources, and physical ones (Table 6-1).

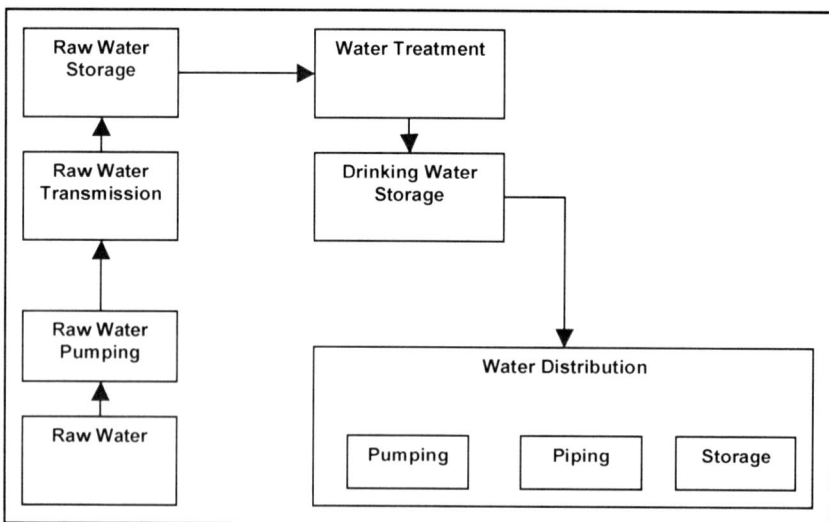

Figure 6-1. Outline of a conventional drinking water system.

Table 6-1. Performance indicators for drinking water systems (modified from Alegre and Baptista 2003).

OPERATIONAL INDICATORS	QUALITY OF SERVICE INDICATORS
Rehabilitation:	**Service:**
Mains rehabilitation (%/yr)	Pressure of supply adequacy (%)
Mains relining (%/yr)	Water interruptions (%)
Replaced or renewed mains (%/yr)	Interruptions per connection (No./1000 connections)
Replaced or renewed valves (%/yr)	Critical interruptions per connection (No./1000 connections)
Service connection rehabilitation (%/yr)	Population experiencing restrictions to water service (%)
	Days with restrictions on water service (%)
Failures and repairs:	Quality of supplied water (%)
Mains failures (No./100 km/yr)	Aesthetic test compliance (%)
Pipe failures (No./100 km/yr)	Water taste test compliance (%)
Joint failures (No./100 km/yr)	Water colour test compliance (%)
Valve failures (No./100 km/yr)	Microbiological test compliance (%)
Service connection insertion point failures (No./100 km/yr)	
Critical mains failures (No./100 km/yr)	Physical-chemical test compliance (%)
Service connection failures (No./1000 connections/yr)	
Hydrant failures (No./1000 hydrants/yr)	**Customer complaints:**
	Service complaints per connection (No. complaints/ 1000 connections/yr)
Power failures (hours/ pumping station/yr)	Pressure complaints (%)
Active leakage control repairs (No./100 km /yr)	Continuity complaints (%)
Water losses:	Water quality complaints (%)
Water losses (m^3/connection/yr)	Water taste complaints (%)
Real losses (l/connection/ day)	Water colour complaints (%)
Infrastructure leakage index (-)	Interruption complaints (%)
	Critical interruption complaints (%)

Continued on next page.

Table 6-1 continued.

FINANCIAL INDICATORS	WATER RESOURCES INDICATORS
Annual costs:	Inefficiency of use of water resources (%)
Unit total costs ($€/m^3$)	Annual water resources availability ratio (%)
Unit running costs ($€/m^3$)	PHYSICAL INDICATORS
	Transmission and distribution storage capacity (days)
Annual investment for network mains:	Valve density (No./km)
Unit investment for network mains ($€/m^3$)	
Annual investments for new and upgraded mains (%)	
Annual investments for mains replacement (%)	
Tariffs:	
Average water charges for direct consumption ($€/m^3$)	
Average water charges for exported water ($€/m^3$)	

6.2 THE WASTEWATER COLLECTION SYSTEM

The main feature of the wastewater collection system is to drain urban areas and, thereby to prevent the flooding of buildings and surfaces. The drainage of urban areas requires collection systems for both wastewater and stormwater. The wastewater collection system needs to be dimensioned to transport all drinking water returned as wastewater; often it must also transport the drainage water from buildings. The stormwater collection system is dimensioned for a storm of a given statistical return period, usually between 1 and 10 years. To prevent flooding of surfaces, even longer return periods (10 - 30 years) may be chosen.

The most common system built today is the separate system (Figure 6-2); wastewater and stormwater are transported in separate sewers. Especially in older urban areas, however, they are transported together in one sewer, a combined system (Figure 6-3). Since the pipes, for economic reasons, can be dimensioned only for rain events of pre-determined intensity, it is not possible to transport all water all the way through the system during unusually intensive rain events. Therefore, all combined systems are equipped with combined sewer overflows (CSOs), at which the exceeding flow is disposed to the recipient untreated. The purpose of these is to prevent flooding, and to protect the wastewater treatment plant from too high flows.

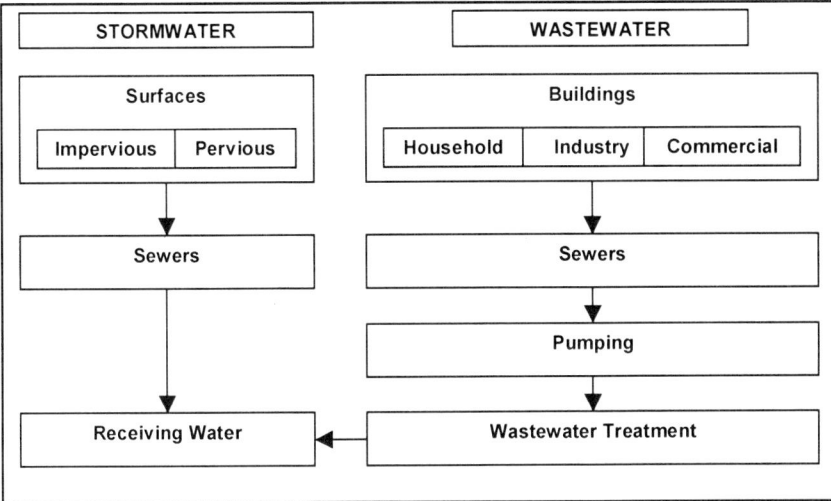

Figure 6-2. Outline of a conventional separate system.

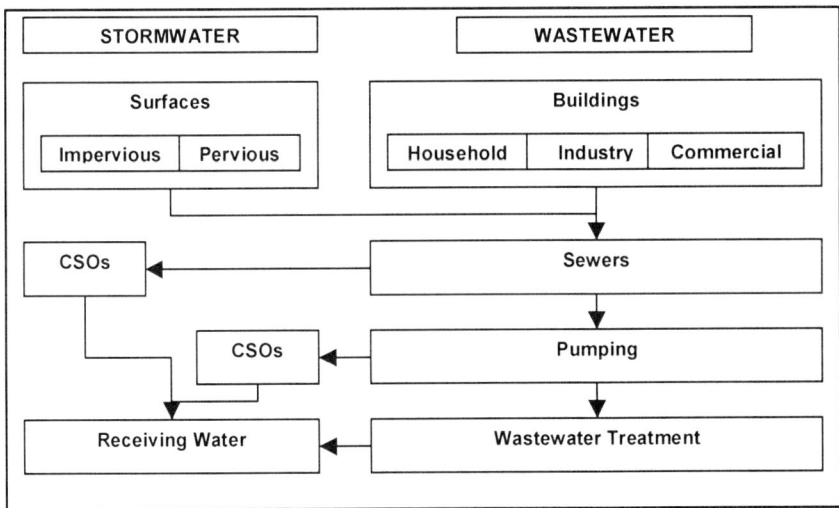

Figure 6-3. Outline of a conventional combined system.

The *Care-S*, a European Fifth Framework Research Programme, has defined a framework for the operation, maintenance and rehabilitation of wastewater collection systems, based on performance indicators (Matos *et al.* 2003a;

2003b). The indicators are divided into: environmental indicators, operational indicators, the quality of service, economic and financial ones, and physical ones (Table 6-2).

Table 6-2. Performance indicators for wastewater collection systems (modified from Matos *et al.* 2003b).

ENVIRONMENTAL INDICATORS	QUALITY OF SERVICE INDICATORS
Wastewater:	**Flooding:**
Overflow discharge frequency (No./overflow device/yr)	Flooding affecting properties: from sanitary sewers in dry weather (No./1000 properties/yr)
Overflow discharge volume (m^3/overflow device/yr)	Flooding affecting properties: from combined sewers in dry weather (No./1000 properties/yr)
Duration of overflow discharge (hours/overflow device/yr)	Flooding affecting properties: from sanitary sewers in wet weather (No./1000 properties/yr)
Overflow discharge related to rainfall (%/yr)	Flooding affecting properties: from combined sewers in wet weather (No./1000 properties/yr)
Sediments:	Surface water flooding of properties: in wet weather (No./1000 properties/yr)
Sediments from sewers (ton/km sewer/yr)	**Interruptions:**
	Interruption of wastewater collection and
OPERATIONAL INDICATORS	transport services (%)
Sewer Cleaning:	**Complaints:**
Sewer cleaning (%/yr)	Blockage complaints (No. /1000 inhab./yr)
Sewer Rehabilitation:	Flooding complaints (No. /1000 inhab./yr)
Sewer rehabilitation (%/yr)	Pollution incident complaints (No. /1000 inhab./yr)
Sewer renovation (%/yr)	Odour complaints (No. /1000 inhab./yr)
Sewer replacement or renewal (%/yr)	ECONOMIC AND FINANCIAL INDICATORS
Manhole chambers replacement, renewal, renovation or repair (%/yr)	**Costs:**
Service connection rehabilitation (%/yr)	Unit total cost per length of sewer (€/km sewer/yr)

Continued on next page.

Table 6-2 continued.

OPERATIONAL INDICATORS *continued*	ECONOMIC AND FINANCIAL INDICATORS *continued*
Inflow, Infiltration & Exfiltration:*	Unit running cost per length of sewer (€/km sewer/yr)
Inflow (m^3/km/yr)	Unit running cost for maintenance, cleaning and repair per length of sewer (€/km sewer/yr)
Infiltration (m^3/km/yr)	**Investment:**
Exfiltration(m^3/km/yr)	Unit investment (€/km sewer/yr)
Inflow / Infiltration / Exfiltration (I/I/E) (%)	Investments for new assets and reinforcement of existing assets (%)
	Investments for asset replacement and renovation (%)
Failures:	
Sewer blockages (No./100 km sewer/yr)	PHYSICAL INDICATORS
Sewer blockage locations (No./100 km sewer/yr)	Sewers
Repeat sewer blockage locations (No./100 km sewer/yr)	Surcharging** in gravity sewers in dry weather (%)
Pumping station blockages (No./100 km sewer/yr)	Surcharging in gravity sewers in wet weather (%)
Surface flooding (No./100 km sewer/yr)	High sewer surcharging (%)
Sewer collapses (No./100 km sewer/yr)	

*   Infiltration: Seepage of groundwater into the sewer.
     Exfiltration: Leakage of wastewater from the sewer.
**   Surcharging: Occurrence of pressurised flow in gravity sewers.

6.3 TECHNICAL FUNCTION FOR FUTURE SYSTEMS

Introducing new technical solutions, or using existing systems in a different way means that new risks are introduced. When alternative urban water systems are developed to meet the requirements of sustainability, they also need to live up to the requirements of technical functionality, the performance indicators set out by the Care-W and Care-S programmes can evaluate this.

In the Urban Water programme, Olofsson B. *et al.* (2001) developed a risk assessment method, based on established risk models (Grimvall *et al.* 2003). Not only were risks assessed in the category technical function, but also *environmental risks* and *health risks*. Robustness, a measurement of a system's ability to resist temporary disturbance from the outside, was estimated in a way similar to that of risk. The risk model was applied and refined in some of the Urban Water model cities.

In the *new city area* study (Hammarby Sjöstad, Section 9.1), it was found that the source separating systems were as good as the conventional wastewater system for the same risk events. The source separating systems, however, introduced additional risk events, because the wastewater is separated into more fractions than in the conventional system. It was also found that the stormwater system has fewer risk events than either the drinking water system or the waste water system, especially regarding environmental and health risks

In the *city centre* study (Vasastaden, Section 9.5), it was found that the existing combined system had the lowest technical risks. The two alternative systems, *i.e.* a separated one with source control, and a blackwater system, both had equally higher technical risks. The differences in robustness were minor. Also for Vasastaden, hypothetical alternatives for drinking water treatment and supply were analysed (see Section 7). A decentralised system, which supplies untreated raw water and has powerful microbial barriers close to the consumer, may decrease the infection risk. However, care should be taken to evaluate all risk events that may arise when more than one quality of water is to be used in the household. For showering, for example, drinking water quality is obligatory (Section 7; Westrell *et al.* 2002).

A general conclusion from these studies is that the conventional large-scale urban water systems have a more reliable technical function. Many small-scale wastewater systems also expose the users to more health-related risk events, and also to higher cumulative risks.

7

Drinking water treatment and supply

Gerald Heinicke and Thor Axel Stenström

7.1 INTRODUCTION

On the global scale, the access to safe drinking water is essential to health, a basic human right and a component of effective policy for health protection (WHO 2004). The UN General Assembly has declared the years 2005 to 2015 as the International Decade for Action, "Water for Life", and clean drinking water has been designated a priority of the Millennium Development Goals.

The challenges to achieve this and distribute safe drinking water vary widely in the world. This is due to the availability, abundance and quality of water resources, the existing technical infrastructure and organisations, as well as lifestyles. In the industrialised countries, waterworks have typically been designed thirty or more years ago and are now facing new challenges. These are partly based on more stringent regulations, as well as the discovery of new chemical and microbial threats, variations in raw water quality, changing water consumption patterns, demands for sustainable production, and high expectations from the consumers regarding the aesthetic quality of tap water.

The new and revised WHO guidelines for drinking water quality (WHO 2004) emphasise the importance of risk assessment and management. This approach is based on the Stockholm Framework (Bartram *et al.* 2001), which advocates holistic safety management of microbial and chemical constituents, from the catchment to the consumers. Other emerging questions are related to the variability in raw water quality and treatment efficiency and to the vulnerability of the exposed population. Outbreaks of waterborne disease have occurred although the conventional water treatment was functioning correctly according to traditional operation criteria. In this context, the barrier efficiency of treatment processes against specific pathogenic microorganisms in surface waters is important. Especially for vulnerable individuals, also chemicals, either of biogenic origin, *e.g.* endotoxins (Anderson *et al.* 2002), or algal toxins (Chorus and Bartram 1999), or of anthropogenic origin, *e.g.* drug residuals, may also cause concern. Variability is also reflected by seasonal problems, such as the precipitation of dissolved iron and manganese in the distribution network and the occurrence of earthy-musty odour, which are causes of consumer complaints. A perceived microbial risk, an impaired aesthetic quality, or interruption of distribution may seriously undermine consumer trust in tap water.

During the past decades, changes in surface water quality have been observed in large parts of central and northern Europe, primarily as increased colour, caused by humic substances. Many conventional waterworks reacted by increasing chemical dosages or reconsidering their treatment options (Nordtest 2003). Whether this trend will continue is debated among researchers. Technical development has provided new treatment alternatives, for example membrane filtration as an economically feasible option that can substantially reduce the content of organic matter and microbial hazards. With the need for renovation at hand, at least in the medium term, planners need up-to date information to find a suitable strategy for their water supply and treatment.

In the Urban Water programme, civil engineers and microbiologists have approached the field of drinking water supply, treatment and distribution cooperatively. This chapter concentrates on investigations of treatment, distribution and microbial risk assessment. In a systems analytical study of centralised versus decentralised treatment, environmental issues were also included. In addition, the projects have aimed to assess the biological role of biofilms in drinking water treatment and distribution.

7.2 SUSTAINABLE DRINKING WATER TREATMENT

Sustainability may be defined from several angles: technical choices, production alternatives, social acceptance, cost-benefit, health impact or environmental considerations.

Traditionally, the sustainability of urban water systems has focused on energy consumption and the use of resources. Energy consumption, for Europe in generally, is around 4700 W per person, which according to Imboden (2000) should be reduced to around 2000 W as a sustainable level. What percentage of this is reflected by the drinking water production? Based on a life cycle analysis of the production of drinking water at Lackarebäck waterworks in Göteborg, Wallén (1999) concluded that the environmental burden can be mainly attributed to energy consumption. In Göteborg, taking into account the water production and supply, including chemicals, road transport, maintenance of the pipe network, and a total production of 345 litres per person and day, the result is equivalent to a continuous consumption of 7 W per person. Sixty percent of the energy consumption is due to the pumping of water, and therefore remains unaffected by changes in the treatment process. Thus, the current method of supplying water to industry and the households accounts for less than 1%, in a "sustainable society" with a primary energy consumption of 2000 W per person. It is concluded that this level of energy consumption is acceptable for the provision of clean drinking water.

Conventional urban water systems with centralised treatment of drinking-, storm-, and wastewater have been criticised for wasting resources. One such example is the purification of large volumes of water to drinking standard, while only a small fraction is used for potable purposes. Local treatment of water to the level of purity needed for a given purpose, as well as water reuse, has been advocated. Local reuse is not a new concept. Loll (1892) described an in-house installation, implemented in St. Petersburg, which included the reuse of greywater for toilet flushing, as part of a source-separating toilet system for nutrient recovery in agriculture.

Non-potable reuse of water has more recently been implemented in Japan, Australia and the United States (*e.g.* Okun 2000). Rainwater harvesting, common in many developing countries, has also been considered for cities of industrialised nations to decrease the use of potable water. However, large-scale rainwater harvesting is not generally considered economically or environmentally advantageous in moderate climatic zones that do not have pronounced water shortages (Crettaz *et al.* 1999; Mikkelsen *et al.* 1999), and it may involve hygienic risks (Albrechtsen 2002).

Decentralised treatment is another option; basic quality water designated for household purposes is supplied centrally, while a part of the flow is upgraded to drinking water quality by membrane filtration (*e.g.* Ma *et al.* 1998). However, the current trend in the EU is to concentrate more on holistic, catchments-to-tap improvement of centralised water supply systems (McCann 2004). Localised drinking water treatment may act as an additional barrier to pathogens that originate from the raw water or the distribution network (Payment 1998), although the issue of regrowth in the in-house installation has been raised.

In theoretical studies of Göteborg, it was concluded that decentralised treatment with membranes, in terms of both environmental impact (by material flow analysis, MFA, see Section 2) and health (by microbial risk assessment, MRA, see Section 3.1), was a competitive alternative to the centralised conventional treatment and distribution. The aspects studied were energy consumption and microbial risk, based on two scenarios in which all decentralised water treatment takes place within a block of flats (Figure 7-1), and only raw water is to be supplied centrally. Treatment was assumed to be with a 10 nm pore size ultrafiltration (UF) membrane, and the permeate (filtered water) was to be used for all purposes in the household. Alternatively, the complete flow would pass a microfiltration (MF) membrane, while a smaller volume would be further treated, by point-of-use reverse osmosis (RO) units, for drinking water and food preparation purposes. The issues of costs, membrane fouling and pre-treatment, as well as monitoring and maintenance were not addressed quantitatively.

A decentralisation system with local treatment in the buildings was not found, theoretically, to have much effect on the energy consumption. The microbial risk assessment indicated that integrating membrane filtration in the process brought the infection risks below the acceptable risk level (10^{-4} infections per person and year), provided the filtered water was also used in showering. Technically, membrane filtration of surface water requires pre-treatment to control fouling. Furthermore, integrity monitoring and maintenance would have to be established for a large number of units. Semi-decentralised systems that treat drinking water close to the consumer, but for a larger number of households, may overcome some of the above problems. Local drinking water treatment may offer a more robust infrastructure, in the sense that varying potential amounts of pathogens in the raw water or sewage ingress and other contamination that may occur within the distribution network would be counteracted by the treatment at the point of use.

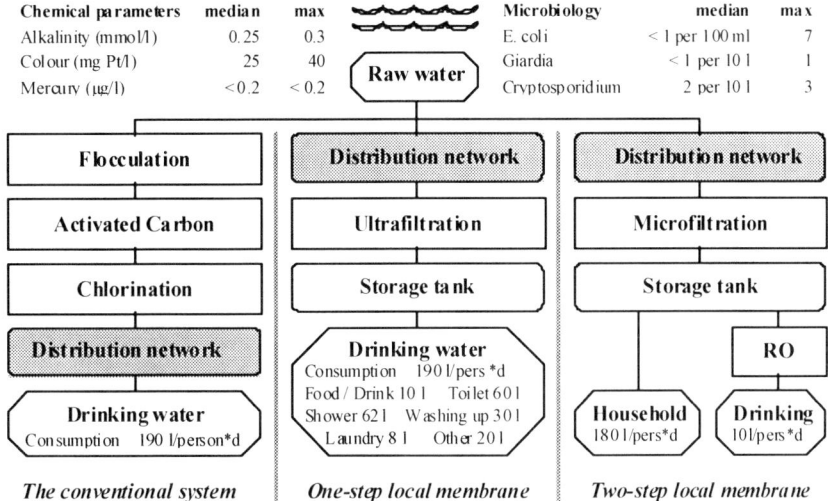

Figure 7-1. Flowchart of the conventional drinking water system and alternatives for local membrane treatment in one step and in two steps (Westrell *et al.* 2002). Year 2000 water quality data from Lake Delsjön, Göteborg.

7.3. INVESTIGATIONS OF PROCESS COMBINATIONS FOR SURFACE WATER TREATMENT

Investigations of treatment processes were conducted at Lackarebäck waterworks in Göteborg, which treats soft, moderately humic surface water of low turbidity. Both in its raw water composition and processes, the waterworks is a typical example of surface water treatment in Sweden. The process studies were conducted on pilot scale (Figure 7-2). A pilot plant at the site was operated to closely resemble the full-scale treatment, *i.e.* conventional coagulation, sedimentation, and rapid media filtration (Heinicke *et al.* 2006). Nanofiltration was also investigated, and biological pre-treatment evaluated for both separation techniques. The feed water to the pilot coagulation train was alternated weekly between raw water and biofilter effluent, to achieve a quasi-paired data set and to counteract seasonal influences on the comparison of treatment efficiency with, and without, pre-filtration. The investigations cover more than one year, thereby accounting for the annual variation of water temperature and quality (Heinicke 2005; Heinicke *et al.* 2006).

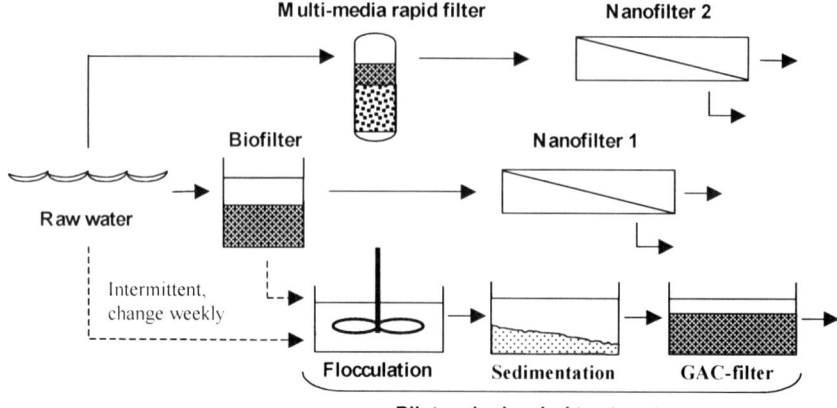

Figure 7-2. Schematic of pilot plant set-up with biofilter, multi-media rapid filter, conventional coagulation & filtration treatment, and nanofilters (Heinicke 2005).

7.3.1 Biological pre-filtration

Biofilters were operated with either granular activated carbon (GAC) that was exhausted for the adsorption of natural organic matter (NOM), or with crushed expanded clay (EC). The filters were fed raw water at an empty-bed contact time of around 30 minutes, and backwashed when necessary. Ozonation, which may be used to increase the biodegradability of NOM by partial oxidation, was not used in this study, to avoid introducing further biological instability. While the removal of bulk NOM (measured as total organic carbon or UV-absorption) was low, about 30% of the biodegradable fraction was removed. The biofilm formation potential in biofiltrate was reduced by 80 - 90% compared with raw water (Persson *et al.* 2005a).

The counting of autofluorescent particles (microalgae) by flow cytometry was introduced, by Bergstedt and Rydberg, (2002) as a tool to monitor the microbial barrier function in water treatment. Since autofluorescent particles are not produced during treatment, they are indicators for the removal of biological particles that originate from the raw water. Two size intervals were monitored: 0.4 - 1 µm (bacterial size) and 1 - 15 µm (protozoan size range). The removal of algae (Akiba *et al.* 2002) and autofluorescent particles 1 - 15µm (Bergstedt and Rydberg 2002) has been shown to correlate with Cryptosporidium parvum removal.

The removal by biofiltration of all the particle fractions investigated was 60 to 90%. The parameter "total particles" was removed to a greater extent than that of autofluorescent particles. The biofilters were also challenged with 1 μm fluorescent microspheres, with hydrophobic and hydrophilic surface characteristics, and bacteriophages (viruses that attack bacteria). The initial removal of the added microspheres was removed at 97 - 99% (hydrophobic) and 84 - 89% (hydrophilic) after 3 hydraulic residence times (HRT); microspheres retained in the biofilter media slowly detached into the filtrate weeks to months after being added. The removal of bacteriophages (after 3 HRT) was considerably lower (40 - 61%), and no further detachment was observed. A comparison of experimental data with theoretical predictions for removal of particles in media filters (clean bed conditions) revealed a similar or greater removal for particles around 1 μm in size than predicted, while bacteriophages were removed at a similar or lesser extent than predicted. The results highlight the selectivity of the particle removal processes; they also have implications for the operation and microbial risk assessment of a treatment train with biofilters as a pre-treatment (Persson *et al.* 2005b).

Earthy-musty taste and odour, difficult to remove by treatment processes, is a widespread problem in surface water supplies. Two compounds (geosmin and MIB) known to cause odour episodes in surface water were effectively removed (by >80%, or below the detection limit) by direct biofiltration of surface water. This was better than for the complete conventional full-scale process with coagulation and GAC rapid filtration. The removal mechanism, *i.e.* biodegradation versus adsorption, was investigated for biofilters with semi-exhausted GAC and non-adsorptive EC media. For the EC, the removal was entirely through biological activity, while for GAC an adsorption capacity remained when the metabolic activity was suppressed. This means that even GAC, which has already been used for several years for the pre-treatment of surface water, added some robustness to the removal of hydrophobic and biodegradable trace organics, which occur intermittently in the raw water (Persson *et al.* 2005c, Heinicke *et al.* 2006).

Biological pre-filtration is particularly interesting for the treatment of raw waters subject to episodes of taste and odour or of dissolved iron and manganese species. Furthermore, it offers – as a sort of riverbank filtration "light" – an additional barrier to peak loads of particles and NOM in surface waters. Integrating a new process into an existing conventional treatment train implies considerable investment costs. In each specific case, the benefits need to be weighed against cost and compared with other options of process upgrade, such as ozonation-biofiltration, UF for increased particle removal, or nanofiltration.

7.3.2 Conventional coagulation treatment

A treatment train consisting of coagulation, flocculation, sedimentation and rapid media filtration is the most common way of treating surface waters. This process removes natural organic matter (NOM) that causes colour, as well as particles and microorganisms. The barrier function of conventional coagulation and filtration was evaluated for particles and microorganisms, by weekly sampling, on-line particle counts, and challenge tests. During normal operation, coagulation and filtration, for both pilot study and full-scale operation, achieved a 2-log removal of raw water derived particles in the size range 0.4 - 1 µm (bacteria size) and a 1-log removal in the 1 - 15 µm range (Heinicke *et al.* 2006).

The local records of incidents in the treatment process showed that a malfunction in coagulant dosage was the most common type of failure. To assess the consequence of a disturbed coagulant dosage, challenge tests with the addition of 2% wastewater (to increase the microbial load for reduction measurements) were conducted with varying coagulant doses in the pilot plant. At a sub-optimal coagulant dose, the removal of particles by flocculation and sedimentation deteriorated, which increased the load to the rapid media filter and, subsequently, resulted in higher numbers of faecal indicator bacteria in the first filtrate after backwash. Figure 7-3 shows the time-series of a simulated disturbance of coagulant dosage lasting 24 hours. The vertical lines mark the beginning and end of the lowered dosage (31% of normal). The formation of visible floc ceased, as did the effect of flocculation and sedimentation on the concentrations of coliform bacteria (Johansson, A. and Scott 2004).

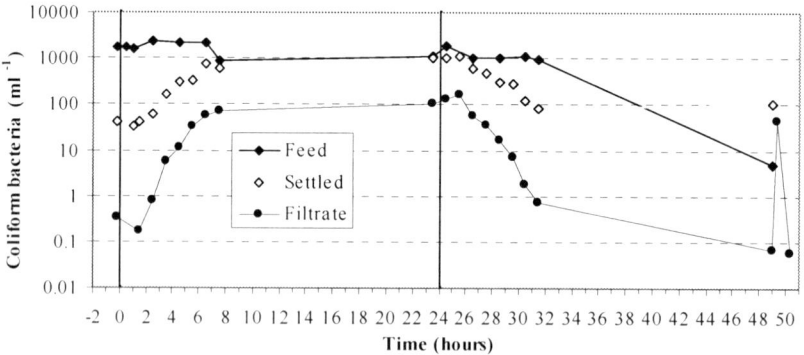

Figure 7-3. Coliform bacteria in the chemical treatment pilot plant during a simulated disturbance of coagulant dosage. Most probable number by Colilert-18™. Vertical lines delimit the beginning and end of the phase with 0.81 mg/l Al dosage (31% of normal). The last three data points in the filtrate are just prior to a backwash, directly after it, and one hour after it. Modified from Johansson, A. and Scott (2004).

The efficacy of coagulation and filtration for virus removal was assessed in the pilot plant with a challenge test using bacteriophages. The aims were to assess the relative impact of the subsequent treatment steps and to compare the reduction of two bacteriophage types commonly used as surrogates for viral reduction (Heinicke *et al.* 2004). Bacteriophages of the types φX174 and MS-2 were added either before the addition of chemicals or in the first flocculation chamber. The results are summarised in Figure 7-4. The concentrations of φX174 were less reduced than those of MS-2; hence the former constitutes a more conservative indicator of the barrier function. The chemical addition in itself may result in a more than 1-log reduction. This will probably vary due to the type of coagulant and other conditions. The flocculation will result in an additional log reduction, as will the sedimentation/filtration. The total reduction was in the range of 3.8 logs. Further investigations are needed to evaluate and optimise coagulation treatment in relation to virus removal.

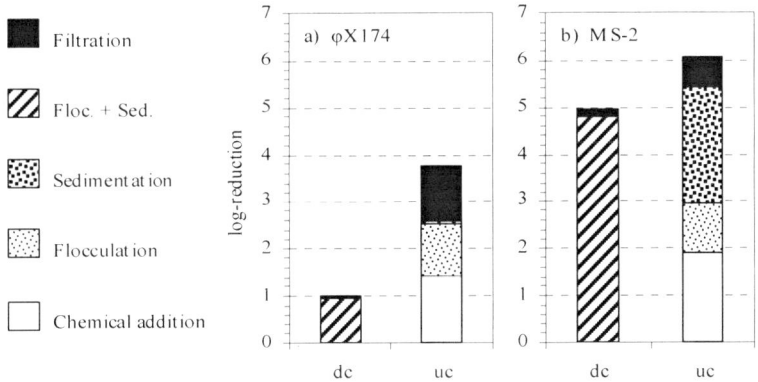

Figure 7-4. Log-reduction of: a) φX174, and b) MS-2 bacteriophages during treatment. The dc = downstream case, bacteriophage addition into first flocculation chamber; for this, combined reduction over flocculation and sedimentation. The uc = upstream case, bacteriophage addition upstream of chemical addition, differentiated between flocculation and sedimentation (Heinicke *et al.* 2004).

Coagulation, sedimentation and rapid media filtration constituted a partial barrier to particles in the size range of pathogenic bacteria and parasitic protozoa, although the barrier was robust in resisting changes in process parameters. This robustness contradicts reported failures of barrier function under even slightly sub-optimal conditions (Emelko 2003); it may be due to the design of the treatment train investigated, with long retention times.

The mediocre barrier function for raw water derived particles under normal operating conditions may make it necessary to improve particle removal. This may hold true especially for waterworks with a rapid variation in the concentration of particles and natural organic matter (NOM) in their raw water (Hurst *et al.* 2004). If an additional powerful particle barrier were included in the process train (*e.g.* ultrafiltration), the flocculation process could be optimised for the removal of NOM instead.

7.3.3 Nanofiltration

Nanofiltration (NF) has emerged in recent years as an alternative to conventional coagulation and filtration. The main drawback of the technology is its susceptibility to membrane fouling, *i.e.* a partly irreversible decline in permeability. Two NF pilot plants were operated in parallel, at a comparatively low flux (hydraulic load). This approach, with simple pre-treatment, such as rapid sand filtration, is established mainly in Norway. One plant (NF1) received biofiltrate, while the other (NF2) received multi-media rapid filtered feed water (Figure 7-2). The polyamide thin-film composite membrane was designed for high NOM removal and salt passage.

During the 19 months of operation, both plants exhibited fouling. Especially in the membrane elements that were fed rapid filtered water, a rapid development of trans-membrane pressure occurred, and the permeability could not be fully restored by alkaline cleaning. Cleaning intervals of less than one month and membrane life times below 3 years are not considered economically feasible. Thus, it was concluded that the above-specified membrane was not feasible for surface water with rapid filter pre-treatment only. The pressure drop was, however, significantly mediated by biological pre-filtration (Persson *et al.* 2005d).

After the operational phase, pieces of the membrane and its biofilm were sampled and analysed for chemical and microbiological parameters. The analysis of the fouling layer did not reveal a primary cause of the difference in pressure development; all parameters analysed were about a factor of 2 higher on the membrane with rapid filtration pre-treatment. It is probably a combination of the lower concentrations of particles and biodegradable organic matter in the biofilter that contributes to the improved function. Biopolymers, *i.e.* organic compounds of high molecular weight that are produced by microorganisms, have been identified as one of the reasons for membrane biofouling (Persson *et al.* 2005d). Figure 7-5 illustrates the differences in the fouling layer.

Figure 7-5. Photograph of the feed side of the NF membrane, with the feed spacer. From left to right: Operated on biofiltrate, clean membrane, operated on rapid filtrate.

7.4 BIOFILM IN THE DISTRIBUTION NETWORK AND THE EFFECT OF PRIMARY DISINFECTANTS

In a section of the Stockholm Water distribution network, the effect of two primary disinfection methods (UV treatment and chlorination) was investigated for biofilm development and assessed by *in situ* flow cells (Långmark 2004), also taking into account seasonal variations. As expected, both the heterotrophic plate counts and the total number of viable bacteria, assessed microscopically, increased in the biofilm with rising residence time. No significant differences were seen between the parallel lines.

The pattern of free-living protozoan "grazers" was also similar for corresponding sites in the full-scale systems. The grazers seem to be of primary importance in controlling the biofilm within the distribution systems. Two 1000 m long pilot networks were built and fed with the same water as the full-scale network. For the same residence time, the biofilm development in the pilot network was similar to the full-scale one. The pilot networks were challenged with 1 μm fluorescent microspheres (as a standard particle), fluorescence-marked Legionella (as a bacterial model) and bacteriophages (Långmark *et al.* 2005). The accumulation, persistence and fate of the added particles and organisms were measured for 38 days to evaluate the effectiveness of the biofilm. The accumulation of hydrophobic spheres was in the range of 0.1 - 0.8% (10 - 100 particles/cm); the hydrophilic ones were accumulated by at least one magnitude more, probably due to easier penetration of and subsequent entrapment in the biofilm. The fluorescence-marked Legionella followed essentially the same pattern in the parts of the network closest to the waterworks, and showed a pattern nearly identical to that of the conservative model fluorescent particles. The bacteriophage accumulation was low; it represented about 0.001% of those introduced. The detachment of fluorescent microspheres over the course of the experimental period did not differ for the types of disinfectants applied. The detachment of bacteriophages was, however, significantly higher in the UV treated system. Detachment from the biofilm occurred under both laminar and turbulent flow conditions; in the laminar regime continuously, and in the turbulent regime as larger aggregates. The investigations, which gave a lower than expected accumulation of organisms within the biofilm, but a higher persistence, indicate the potential of using fluorescent microspheres as a conservative model particle in similar studies.

7.5 MICROBIAL RISK ASSESSMENT

Questions of variability are related to (1) the barrier efficiency exemplified above, (2) fluctuations in the raw water quality, (3) frequency and severity of failures and incidences both in the waterworks and in the distribution network, and (4) variations in consumer consumption of drinking water. The relevance of such variability, which can be investigated by quantitative microbial risk assessment (QMRA), has also been addressed within the Urban Water framework. In an international cooperative investigation about the seasonal fluctuation of Noroviruses in raw water (Westrell 2004), it was shown that there was a distinct winter peak, revealed during an intensive sampling campaign. The occurrence probably reflects outbreaks and seasonally high prevalence within the population. Elevated numbers of Noroviruses corresponded to increased turbidity, which probably reflects heavy rainfalls with combined sewer overflows.

Data on the failure incidence rates in water treatment and in the distribution network in Göteborg were assessed. Calculated on a yearly basis, a larger number of people were estimated to be infected by pathogens not removed by the treatment under normal operating conditions, than by failure incidents. The incidents in the waterworks were short (Table 7-1) and did not have a substantial effect on the yearly risk of infection; however, the potential of plug-flows was not fully accounted for. Incidents in the distribution network were less common and affected only a few people at a time, however these people were at high risk of becoming infected (Westrell *et al.* 2003).

Rotavirus, used as a model for viral transmission, caused the largest number of potential infections. Most infections from Campylobacter probably originate from contamination of reservoirs in the distribution network. The calculated risk of Cryptosporidium infections was significantly lower than for the other two pathogens. A sensitivity analysis pointed out the concentration of pathogens in the raw water, the chlorination (for Campylobacter), and the dose-response functions to be the most critical factors in the risk of infection. The total number of simulated annual infections in the system was within the range of the fraction of gastroenteritis attributable to tap water estimated from epidemiological data.

Table 7-1. Risk incidents identified in the Göteborg study, adjusted to account for the entire waterworks and distribution network (from Westrell *et al.* 2003).

Type of incident	Consequence	Frequency[1]	Duration[2]
Sub-optimal removal of particles in drinking water treatment			
- Wrong dosage of coagulant, or pH for precipitation	No removal of pathogens	1.5	0.6
- Disturbed filter operation	No removal of pathogens	15	5
Chlorination failure in drinking water treatment	No post-chlorination	0.5	0.4
Pollution of water during distribution			
- Cross-connections	1% sewage	0.00004 (or 0.4 affecting 25 people)	3 days
- Pollution entering part of system without pressure storage tank or water main	0.002% sewage	0.007 (or 0.7 affecting 2500 people	14 days
- Service pipe or periphery network	0.002% sewage	0.00003 (or 0.3 affecting 25 people	30 days

[1] Incidents per year and affecting 250 000 people; [2] Hours per incident

The variation in consumption patterns was investigated for the Swedish population (Westrell *et al.* 2006). The water consumption is related to demographic and socio-economic factors. Women consumed larger quantities of cold tap water than men (on average 0.95 litres compared to 0.79 litres), and the consumption decreased with increasing income (Westrell 2004). Sensitive subpopulations, such as elderly people, had a higher consumption of cold water than other groups. Bottled water consumption was generally low.

There are indications that conventional treatment of surface water may not always be sufficient to keep infection rates below internationally agreed limits, *i.e.* 10^{-4} infections per person and year or 10^{-6} DALYs[1] (Westrell *et al.* 2003). Possible solutions are to include additional barriers, such as UF, at the waterworks. In addition, slow sand filtration and disinfection by UV irradiation constitute barriers, although not for all groups of pathogens alike. The MRA tool is described in detail in Section 3.1.

7.6 CONCLUSIONS AND RECOMMENDATIONS

It can be concluded that the different types of variability greatly influence the barrier performance of treatment processes, as well as the aesthetic and hygienic quality of the water. Questions related to failure incidents in both the waterworks and in the distribution network have not received enough attention in the water industry: they should be addressed more explicitly in the future. Challenge tests, here exemplified with bacteriophages and fluorescent microparticles, can be useful to assess failure incidents both in the treatment and in the distribution of water.

Surrogate particles that are detectable on-line, such as microalgae and particles in the size ranges of the microorganisms in question, may be used to assess treatment performance. It was shown for example that the removal of biological particles, in the relevant bacterial and protozoan size ranges, by conventional coagulation and filtration of surface water was only mediocre, which may result in sporadic cases of waterborne disease that are difficult to detect. Although the results need to be confirmed by additional case studies and the available in-data refined, an improved particle removal in surface water treatment can be advocated for. Furthermore, challenge tests, here exemplified with bacteriophages and fluorescent microparticles may play a role in assessing events in both

---

[1] DALY = disability adjusted life year. Measure of the damage to a person's health, here by waterborne disease.

the treatment and the distribution. Microbial risk assessment is a powerful predictive tool, but the results need to be validated by further studies. The method has already started to become a reliable base for decision-making.

8

Integration of complex knowledge

*Jaan-Henrik Kain, Erik Kärrman and
Henriette Söderberg*

8.1 INTRODUCTION

For water utilities, sustainable development represents new and complex planning situations. Comprehensive assessment of alternatives for sustainable urban water management is a multi-dimensional challenge. Incomparable kinds of information need to be used in the same situation, including various types of lay, professional and scientific.

Methods have to be developed with consideration for the intricacy that concepts such as sustainable development, urban water systems and knowledge represent when scrutinised further (Kain 2003; Söderberg and Kärrman 2003). Planning and decision-making processes should also take into account the issue of participation. After all, it is actors who need to integrate knowledge in order to make sensible

© IWA Publishing 2006. *Strategic Planning of Sustainable Urban Water Management* edited by Per-Arne Malmqvist, Gerald Heinicke, Erik Kärrman, Thor Axel Stenström and Gilbert Svensson. ISBN: 1843391058. Published by IWA Publishing, London, UK.

decisions. Consequently, methodologies usable and supportive for stakeholders and decision-makers should not be confused with tools that function as decision machines, *i.e.* where data is inserted into computer software that, subsequently, deliver apparently unanimous answers to the problem at hand.

Managing the planning of water utilities, thus, necessitates competence to work with and to guide groups of stakeholders of all categories. In addition to skills in facilitating group work, there is a critical need to integrate diverse elements of information in order to construct a relevant and coherent foundation for decision-making. It is important to note that this synthesis takes place at the interface between several forms of knowledge, practical, tacit, lay, expert and scientific, which need to be taken into account (Söderberg and Kain 2006). Consequently, there is great potential in methods that facilitate the evaluation of strategies for urban water development across multiple fields of knowledge.

One category of such methods is the Multi-Criteria Decision Aids (MCDAs). In this chapter, an evaluation of some MCDA methods is described, and recommendations based upon these results are brought forward. The themes, described in more detail below, used to structure the evaluation are: symmetrical management of different forms of knowledge; management of heterogeneity, pluralism and conflict; functionality and ease of use; as well as transparency and trust. As can be seen from these themes, the focus is on MCDAs as actor systems, not as expert systems.

8.1.1 Multi-Criteria decision aids

Friend and Hickling (2005) have structured the planning and decision-making part of the policy process as consisting of four interrelated and entwined modes:

- In the *shaping mode*, the problem is structured, differing views of the problem are expressed and discussed, and a common understanding of the situation is beginning to emerge.
- During the *designing mode*, alternative solutions for handling the problem are defined.
- In the *comparing mode* these alternatives are evaluated in relation to different knowledge areas
- The *choosing mode* concludes in an agreement on which alternative the group prefers.

This chapter focuses on one of these phases, the comparing mode, where alternative solutions are assessed with relevant criteria (Friend and Hickling 2005). Although we have been working more or less with all phases in case studies, the comparing mode is the one phase of strategic interest in this context.

Since the criteria used for comparative assessments are multifaceted, and sometimes incommensurable, comparing them is demanding in terms of management of complexity and uncertainty, hence the need for MCDA tools.

There exist many approaches, which, in opposition to single criterion models, such as Cost Benefit Analysis, have developed multi-criteria models (van Moeffaert 2003a). Many of these MCDAs are deterministic distance-to-target procedures, of less use when targets and preferences are not (easily) agreed upon – situations typical for multi-stakeholder processes (Söderberg 2002). Of more interest are outranking methods where stakeholder preferences may be expressed in less categorical terms (van Moeffaert 2003a).

Below, three approaches to MCDA will be discussed: REGIME, NAIADE, and STRAD. REGIME and NAIADE are software-based outranking methods that identify the most defendable alternative among proposed strategies (van Moeffaert 2003a). STRAD belongs to a different category of decision support, covering not only the comparing mode, but also the whole cycle of planning and decision-making modes as described above (Friend and Hickling 2005). It is a software version of the original Strategic Choice Approach (Friend and Hickling 2005), providing some added functionality (Stradspan 2005). In terms of MCDA methodology STRAD is rather minimalist while being based on the more straightforward weighted summation approach. Where REGIME and NAIADE are based on pair-wise comparisons of alternatives STRAD employs a simpler linear additive model in which partial assessments are added to provide an overall value. However, STRAD also provides support for displaying degrees of uncertainty related to each assessment.

A critical MCDA issue is how different kinds of criteria are weighted, *i.e.* how their relative importance is expressed. REGIME uses ordinal weights, arranging criteria along a scale from the most important to the least important, but nothing is said about whether one criterion is twice or five times as important as another. STRAD, in contrast, employs a typical cardinal approach by allowing different weights, ranging from insignificant to extreme importance, to be given. NAIADE has no explicit weighting but, during the process of comparison, a perceived small relative difference between the alternative strategies in relation to a specific criterion is interpreted by the software as indicating little weight for that criterion.

8.1.2 Participation, by whom, when and why?

The issue of public participation has gained a renewed interest as a consequence of the political standpoints stated in the Rio documents (UNCED 1993) and successors such as the *'Sixth Environment Action Programme of the European Community'* and, in Sweden, the *'National Strategy for Sustainable*

Development' (European Commission 2001b). The political commitment to a broadened public participation has been remarkable and participation has been associated with the development of an engaged and capable civic body. Beierle and Cayford (2002) have offered an explanation to this renewed interest by arguing that ideas about public participation are best understood as a challenge to traditional management of government policy by experts in administrative agencies. The political aims for increased participation in general, but especially as a component of sustainable development, bring forward a number of theoretical as well as practical problems (Åberg and Söderberg 2003).

In practice the arguments for involving the public in environmental decision-making are varying; moreover, as argued by for example Beierle and Cayford (2002), they are seldom clear. According to the same authors, the purpose of participation has shifted from merely providing accountability to developing the substance of policy. In line with this, arguments found in the literature may range from public participation for generating sustainability indicators to stakeholder involvement for creating social support. Promoting openness and mitigating misunderstandings, probability for success, likelihood to achieve substansive policy changes, as well as improvement of decision outcomes are other examples of provided reasons. Participation can also be seen as a reaction against the science based decision-making: participation implies a contribution to the policy process of forms of knowledge other than scientific, and it also provides an opportunity for the participants to learn from each other.

To summarise, the aims of participation can be divided into six groups:

- Incorporating stakeholder values into decisions,
- Improving the substantive quality of decisions,
- Resolving conflict among competing interests,
- Building trust in institutions,
- Educating and informing the public, and
- Facilitating implementation of taken decisions.

8.2 A COMPARISON OF SIX CASE STUDIES

In this section, findings from six Swedish case studies are discussed. The common denominators are infrastructural planning, multi-stakeholder participation and the use of MCDA tools.

8.2.1 Urban water in Hammarby Sjöstad, Stockholm (REGIME)

Urban Water researchers first designed two wastewater treatment system structures for comparison. The residential area of Hammarby Sjöstad, Stockholm, was chosen as the setting. In August 2001, three groups were gathered to test the synthesis strategy: 1) representatives of stakeholders in Hammarby Sjöstad or Stockholm, 2) stakeholder representatives at a national level and 3) Urban Water researchers. The overall objective for the groups was to decide which of the system structure they preferred. (Etnier and Söderberg 2002). The case is described more in detail in Section 9.1.

8.2.2 Sewage water and organic waste management in Surahammar (NAIADE)

A few local stakeholders, such as local officials and a farmer, assessed three alternatives for urban wastewater and organic waste management across 15 evaluation areas. The Surahammar case study took place in the autumn 2003 and is described more in detail in Section 9.2.

8.2.3 Snow management in Sundsvall (NAIADE)

The Sundsvall case study, which was carried out in northern Sweden, focused on snow management (*i.e.* frozen stormwater). Here a broad range of local stakeholders participated; for one day, in May 2004, they discussed what alternative they preferred out of those suggested by a group of researchers. The stakeholders were: local officers from the Environmental and Public Health Office, the City Planning Office, and the Land and Real Estate Management Office; politicians from the Land and Real Estate Management Committee; and entrepreneurs responsible for transporting the snow as well as for using the snow in a distant cooling facility. Six alternatives for snow management in the central parts of Sundsvall had been suggested: central snow deposits on land, central snow deposits in water, local snow deposits on land, combination of central and local snow deposits on land, separation of snow based on snow quality coupled with a combination of land and water deposits, and, a similar separation of snow but now solely on land deposits.

8.2.4 Storm water in Vasastan, Göteborg (NAIADE)

The Vasastan case study dealt with sustainable storm water management for the district of Vasastan in the city of Göteborg, a part of the city centre. Here, a group of local officers from various divisions of the municipality and representatives from the County Board met for two days in June 2004. The municipal divisions represented were the Environmental Administration, the Department for Sustainable Water & Waste, Göteborg Water and Sewage Works, Gryaab (the regional sewage works), and the local Traffic and Public Transport Authority. The aim of the meeting was to discuss what alternative they preferred out of the four alternatives suggested by a group of researchers. The four alternatives were: 1) Combined system, transporting mixed domestic wastewater and stormwater to the wastewater treatment plant (WWTP); 2) Duplicate system where stormwater is led to the local recipient and wastewater to the WWTP; 3) Duplicate system with local treatment of stormwater, by which cleaner streams of stormwater are infiltrated locally and more polluted streams are treated in ponds before discharge to the local recipient; and 4) Combined system with separate handling of wastewater from toilets (blackwater), by which the remaining fractions (greywater and stormwater) are treated as in the combined system (alternative 1). The Vasastan case study is described more in detail in Section 9.4.

8.2.5 Peri-urban water in Sandviken, Södertälje (STRAD)

The Sandviken case was about planning for sustainable wastewater systems in a Swedish urban periphery, in the Stockholm region. In this example, 13 inhabitants together with three local officers constituted a working group that discussed relevant system structures for the area, as well as criteria for the comparative assessments of these system alternatives. The local officers represented the Environmental Committee, the City Planning Office as well as the municipal water utility. From spring to early autumn 2004 the working group met five times, for two-and-a-half-hour meetings. The work resulted in a list of twelve relevant criteria and seven potential system structures. Subsequently, the working group assessed these systems by referring to the selected criteria.

8.2.6 Sewage water in Uppsala (STRAD)

In December 2005, four alternative sewage water systems were assessed by a group of researchers and local officers according to 12 criteria. The Uppsala case study is described more in detail in Section 9.5. Below, the analysis of the case study material is carried out as a comparative evaluation for five main themes:

- *Symmetrical management of different forms of knowledge:* Do the MCDAs succeed in managing both global demands and locally specific issues? How is input and processing of different forms of knowledge elements supported?

- *Management of heterogeneity, pluralism and conflict:* How are different, and sometimes conflicting, worldviews accommodated? Do the MCDAs support conflict management?
- *Functionality and ease of use:* How do the MCDAs function within the procedural setting? Are there technical malfunctions? Is it possible to swiftly respond *e.g.* to stakeholder discussions and newly raised issues?
- *Transparency and trust:* Do participants, and process facilitators, understand and trust how the tools reach their conclusions? Are outputs accepted by the participants?
- *Participation in practice:* What kind of aim with participation can different tools and methodologies support?

8.3 RESULTS FROM THE CASE STUDIES

8.3.1 REGIME

Symmetry of knowledge

In Hammarby Sjöstad, participants found the structured group discussions, in themselves, to be an advantage of the MCDA; this is an effective way of putting diverse forms of knowledge on a common 'table' where not only expert and scientific knowledge, but also practical and local knowledge is examined (Söderberg 2002). The information material provided by the research project was not fully appreciated by the participants. Data on environmental and economic aspects were accepted, but information on hygiene, socio-culture and technical function were looked upon with more suspicion. However, this has less to do with the MCDA tool per se, but is related to how we perceive sources and fields of knowledge in general (Söderberg and Kain 2006).

REGIME uses a qualitative interface for the assessment of alternative strategies. This has the clear advantage of placing quantitative and qualitative knowledge input on an equal footing. Even so, it may be a disadvantage that the primary merit of quantitative knowledge, *i.e.* the provision of exact figures of measured phenomena, risks being lost in the translation from quantitative to qualitative idioms.

Heterogeneity and pluralism

Both the assessments of urban water systems and the ranking of the relative importance of different criteria were performed by the participants on an individual basis, for subsequent amalgamation in the REGIME software. In this respect, the tool supports an impartial accommodation of the judgments of all participants. However, many participants found the assessment and weighting procedures difficult since there were no points of reference, *e.g.* "Best possible in relation to what?" (Etnier and Söderberg 2002).

REGIME expressed the process result as a 100% probability that one system was preferable to another. Here the participants protested by arguing that they did not feel such an absolute preference. REGIME thus failed to reflect their uncertainty and mixed preferences (Etnier and Söderberg 2002). Nonetheless, the MCDA process supported consensus on the assessment of the urban water systems (Söderberg 2002). However, even if stakeholders preferred the same alternative, they did so for different reasons, by emphasising different criteria (Etnier and Söderberg 2002).

Functionality and ease of use

The DEFINITE software used to implement REGIME was based on DOS, which is not compatible with all Windows based computers. This is inconvenient, in that it is necessary to bring a compatible laptop to any MCDA workshop. In addition, DOS is obviously not the most user and viewer friendly interface format. Another serious drawback was that the results only from one participant at a time could be open, and that the software had to be restarted in order to exhibit the opinion of the next participant. However, DEFINITE is nowadays available in an updated commercial version, which most possibly eliminates some of the problems described above.

Transparency and trust

Some of the participants stated that it was quicker to make the REGIME calculations in their heads than to wait for the computer; this implies that REGIME is transparent. As discussed above, the REGIME output in the form of 100% probabilities was, however, distrusted.

Participation in practice

As the process was designed with criteria and system alternatives decided beforehand, it was mainly characterised by educating and informing, with potential for *facilitating implementation*. This result is, however, not a consequence of the use of REGIME but of the process design.

8.3.2 NAIADE

Symmetry of knowledge

Participants in Sundsvall and Vasastan confirmed that the structured dialogues are the greatest benefit of MCDAs. As regards the comprehensive knowledge material distributed in advance of workshops, the Surahammar and Sundsvall cases show how this becomes downplayed in favour of the much briefer audio-visual presentations just prior to the processing of each criterion (Söderberg *et al.* 2004). One reason for this is the complexity of the knowledge material, where a number of knowledge areas are provided in a quite compressed format making it difficult to access. Another explanation is lack of time on behalf of the participants; they simply do not have the time to read the material. The actual assessments thus tend to rely on a quite superficial foundation. However, the short presentations also allow the participants to adjust and complement the scientific input with more locally derived knowledge.

NAIADE easily accommodates both qualitative and quantitative aspects when assessing alternate strategies, a feature appreciated by all participants (Figure 8-1). However, how the choice between using qualitative or quantitative expressions may influence the output, was clear neither to the participants nor to the process facilitators.

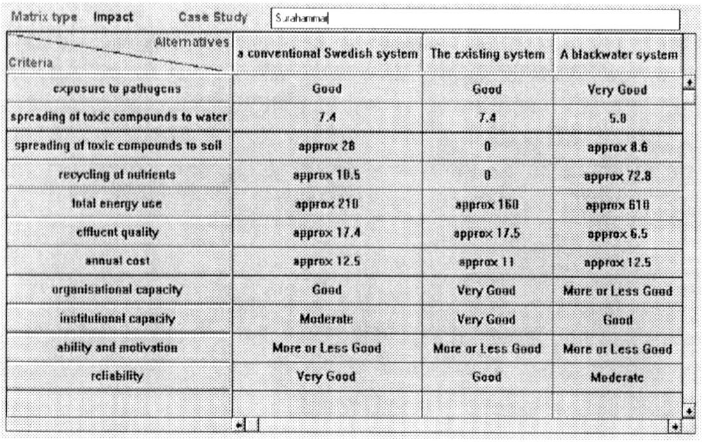

Matrix type Impact Case Study	Sundsvall		
Alternatives Criteria	a conventional Swedish system	The existing system	A blackwater system
exposure to pathogens	Good	Good	Very Good
spreading of toxic compounds to water	7.4	7.4	5.8
spreading of toxic compounds to soil	approx 28	0	approx 8.6
recycling of nutrients	approx 18.5	0	approx 72.8
total energy use	approx 210	approx 160	approx 610
effluent quality	approx 17.4	approx 17.5	approx 6.5
annual cost	approx 12.5	approx 11	approx 12.5
organisational capacity	Good	Very Good	More or Less Good
institutional capacity	Moderate	Very Good	Good
ability and motivation	More or Less Good	More or Less Good	More or Less Good
reliability	Very Good	Good	Moderate

Figure 8-1. The NAIADE evaluation matrix.

Heterogeneity and pluralism

In contrast to REGIME, the NAIADE procedure is based on participants reaching consensus in the assessment of each criterion. Consequently, there is a risk that participants, who for various reasons are more powerful, take control of the assessment process. However, even if the need for temporary consensus may suppress some participants and favour others, the consensus building also serves to unveil different kinds of knowledge and to unravel conflicting opinions.

The indirect weighting procedure was accepted, but without any deeper understanding of its implications (Söderberg *et al.* 2004). In Sundsvall, however, participants would have preferred a more direct weighting to allow them closer control, since they wanted to give more weight to economic and environmental aspects. In Vasastan, in contrast, some participants found direct weighting to be too crude. Another observation is that, since the implicit weighting procedure is active only in relation to criteria expressed in quantitative terms, qualitative criteria run the risk of being either undervalued or overvalued.

NAIADE also includes analysis of potential coalitions among participants. This feature may anticipate future behaviour of stakeholders and predict what strategies have a higher probability of acceptance. However, does knowing of demarcation lines between potential coalitions reduce the level of conflict? Or is it possible that this may reinforce and cement existing disagreement? Moreover, does the lowest common denominator really indicate a good solution for the strategic problem at hand?

Functionality and ease of use

NAIADE runs easily in Windows although some bugs need further attention. It is an MCDA that aims to address uncertainties and divided preferences in decision-making processes. However, the complexity of the software makes it difficult to manage fast enough for an interactive workshop. In the Surahammar case, participants wanted to find the main factors leading to the output, and what alterations of the assessments were necessary to reach another result. Unfortunately, NAIADE could not respond at all to these requests, since it was impossible to easily obtain an assessment of the strategies across only one or a few of the criteria (Söderberg *et al.* 2004). Moreover, NAIADE was not developed as an MCDA for interactive use: its user interface and graphics are not adapted to such procedures. Also, the presentations of the MCDA results are difficult to interpret. It is not clear what alternative strategy is judged as the best, or how much better it is than the second best alternative.

Transparency and trust

It might be expected that for most participants, understanding is a prerequisite for trust. The NAIADE examples, however, indicates the opposite. The participants did not understand much of the software procedures. Even so, participants demonstrated high levels of trust towards the NAIADE results and its application in planning and decision-making processes. It may even be that the complexity and opaqueness of NAIADE generated increasing trust.

Participation in practice

NAIADE was designed for serving four of the six above-mentioned aims of participation, *i.e.* incorporating public values, improving substantive quality, resolving conflicts and facilitating implementation. Among these four, incorporation of public values is an aim that NAIADE can support in theory through the ranking of alternatives; this is, however, only a quite small part of such an incorporation of values. Instead, the results from our cases indicate that the benefit of NAIADE is mainly to improve the quality of both the material that supports the decision as well as the decision output. In terms of resolving conflict and facilitating implementation, these aims were also met, but it was more the process and discussion among the stakeholders than NAIADE itself which promoted these aims.

8.3.3 STRAD

Symmetry of knowledge

Also for STRAD, observations were made regarding the benefits of structured multi-stakeholder discussions. Furthermore, although written material was distributed in Sandviken, most discussions during workshops were based on slide presentations and on the personal experiences of participants. In terms of qualitative and quantitative knowledge input, STRAD facilitated their processing on an equal basis, although the feeding of quantitative knowledge into the software is a bit tricky.

Heterogeneity and pluralism

Also STRAD is a consensus MCDA. In Sandviken, an interesting observation was that one participant, who persistently argued in favour of a dry solution, managed to systematically inch each assessment a little bit in her favour and, thus, to influence the overall rating of the strategies.

STRAD subdivides planning issues into a number of interlinked decision areas. For Sandviken, four of these were defined: organisation, sorting, tank and treatment. The options for handling the decision areas, when linked in alternative ways, resulted in seven potential system solutions. However, in the STRAD procedure this meant that not seven solutions, but in this case 19 options had to be assessed against each comparison area (*i.e.* criterion). The Sandviken evaluation matrix thus became quite complex, even if only some intersections between decision areas and comparison areas were chosen for assessment (*i.e.* the rectangles in Figure 8-2).

Subsequently, each rectangle was assessed by the group of participants, taking knowledge input relevant for each criterion into consideration, and indicating remaining uncertainty or conflict among the participants. The assessment process, in which a group of residents played the major role, went smoothly. However, the subdivision of the assessment process into decision areas and options seemed to induce more problems than advantages. A benefit may be that assessments become more precise when they are more delimited compared to when dealing with an entire system alternative. Nevertheless, in Sandviken the subdivision caused confusion and fragmentation. Consequently, in the subsequent Uppsala case study, each system alternative was treated as one single decision area, which resulted in a less complex evaluation matrix. The disadvantages were minor compared with the experiences of the more complex Sandviken sub-systems approach.

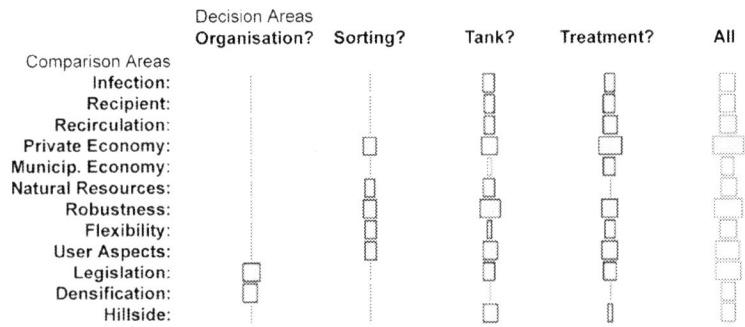

Figure 8-2. The evaluation matrix for Sandviken. Different options within the decision areas are assessed across twelve comparison areas.

The STRAD weighting of criteria is a mix of direct and indirect methods. Direct weighting is done by adjusting the width of the rectangles in the evaluation matrix, where a wider rectangle indicates that more weight is given to that assessment (Figure 8-2). Indirect weighting is done during the assessment of each intersection, where a drawn out scale results in more weight for that particular assessment. This indirect weighting is, however, directly visible in the matrix, as it affects the width of the rectangle. In both Sandviken and Uppsala, none of the participants hesitated to apply weights to the criteria and the "rectangle-approach" to weighting appeared to work well from a user perspective.

Functionality and ease of use

There have been very few signs of software instability, STRAD runs smoothly on any PC. While the graphic interface is more or less acceptable, when compared to REGIME and NAIADE, STRAD is superior. In terms of software flexibility, it is easy to test how different weightings of criteria influence the overall outcome (Kain and Söderberg 2005). In Sandviken, also partial assessments, *e.g.* including only environmental or economic issues, were carried out with ease to illustrate how these criteria affected the overall ranking; this flexibility was much appreciated by the participants.

Transparency and trust

The transparency of STRAD was not questioned in Sandviken and Uppsala. In terms of trust, most participants were positive to working with STRAD again. In Sandviken, where an external consultant was involved, this consultant was so convinced by the output that he decided to abandon his traditional approach to the content of urban water planning in favour of the results from the STRAD session. In Uppsala, many of the participating officials saw potential for STRAD to function both in early stages of formal municipal planning processes and in internal learning processes.

Participation in practice

In the Sandviken example the ambitions for user participation were very high, but the municipality had not given the process results explicit priority over mainstream planning activities. This can be a reason why the process this far has concentrated mainly on facilitating implementation by educating and informing the residents in the working group. A thin line divides the difference between a process focusing on implementation, and a process aiming for incorporation of stakeholder values. To put it a bit crudely, the process facilitator in Sandviken, a consultancy contracted by the municipality, in his working process tended to emphasise paving the way for implementation rather than incorporating stakeholder values. However, based on observations made by the researchers, it became evident that the consultants were not explicitly aware of this. Even so, as with the conclusion drawn for REGIME, this is not a consequence of STRAD but of attitudes of the process facilitator.

8.4 DISCUSSION

So, how do these four MCDAs function within processes of multi-stakeholder interaction? As above, this concluding discussion is structured into four main themes.

8.4.1 Symmetrical management of different kinds of knowledge

In all cases, MCDAs were appreciated in discussions of different categories of knowledge on equal terms, *i.e.* the structured process *per se*. REGIME works with qualitative knowledge input, thus missing out on both the pros and cons of using quantitative knowledge. NAIADE and STRAD allow a mix of quantitative and qualitative knowledge input; where NAIADE has a more precise quantitative interface. The NAIADE approach may lead to an emphasis on quantitative information, since the two kinds of criteria are treated differently in the weighting procedure.

8.4.2 Management of heterogeneity, pluralism and conflict

The REGIME output, even if based on individual assessments, does not reflect remaining uncertainties and mixed preferences. NAIADE and STRAD depend on group consensus for assessment and weighting, which is good for learning but may subdue differences among participants. In terms of weighting procedures, REGIME is straightforward but too crude, while NAIADE is sophisticated but too complicated. The most helpful tool for our case studies was STRAD, with its simple, but still powerful, graphic weighting model. As regards conflict management, the STRAD display of remaining uncertainties is a strength. The coalition feature of NAIADE may be valuable but needs further evaluation.

8.4.3 Functionality and ease of use

The REGIME software was outdated and has been replaced by newer versions. The NAIADE user interface is complicated and there are some software bugs. In addition, its software flexibility is not sufficient. STRAD is the user-friendliest software and is flexible in assessments and weighting. However, all three software packages would benefit from better graphic interfaces.

8.4.4 Transparency and trust

A somewhat alarming observation is that participants display a general trust towards MCDA software, regardless of whether they understand how they work. Even so, there exist some differences. REGIME is transparent but lacks output reliability. The NAIADE software is not transparent, not only to the participants and the process facilitator, but also to our software operator (who had spent significant time in preparation). STRAD is, by far, the most transparent of the four MCDA tools and both participants and process facilitators manage to follow how it is working and trust its outputs.

8.4.5 Participation in practice

In practice, the level of participation in the cases has met the participatory aims of *facilitating implementation* and *improved quality of the decision*. The initial intentions were more ambitious in the Vasastan case, and particularly in the Sandviken case. The aims reached were shown to be heavily dependent on the overall design of the planning and decision-making process, *i.e.* not only the multi-criteria assessment phase. In the three cases where we used NAIADE, the design mainly promoted aims like improved quality of decisions. STRAD is an MCDA with prerequisites to fulfil all the aims of participation presented above. However, the process design and process facilitation skills heavily influence what may be accomplished.

Finally, we would like to emphasise that the future development of MCDA tools would benefit from addressing the following issues:

- An MCDA has to be sufficiently transparent and understandable so that a process facilitator, *i.e.* a non-expert in MCDA software and mathematics, may understand and explain to the participants how the MCDA output is reached.
- Input and equal management of both qualitative and quantitative knowledge is essential to MCDAs.
- A flexibility to try alternative modes of weighting would be of value for better understanding of process outputs among the participants.
- Inherent flexibility of software is essential. Process participants need to examine issues from many angles to analyse the sensitivity of assessments and weighting.
- Graphic and user interfaces of software are of the greatest importance when applied to interactive multi-stakeholder processes. Without good legibility and self-explanatory frameworks, the value of such software diminishes.

8.5 RECOMMENDATIONS

From the perspective of facilitating a process, some recommendations for process design based on our experiences are:

- Agree upon explicit aims with stakeholder participation in the situation at hand. If the aim changes during the process, communicate this.
- The systematic structure of going through one criterion at the time and discussing it within a group of stakeholders was perceived as more important than the aggregation made in the end.
- Match methodology with the aim of the process. It is not always necessary to base the ranking on complex mathematical calculations.
- Do not overkill. Participatory processes should be conducted with care, as they are resource consuming. Select cases thoughtfully and design them with a reflective attitude.

9

Experience from five urban water model city projects

9.1. Daniel Hellström and Marika Palmér Rivera
9.2. Erik Kärrman
9.3. Gilbert Svensson
9.4. Håkan Jönsson
9.5. Erik Kärrman

9.1 A NEW CITY DISTRICT: HAMMARBY SJÖSTAD

Hammarby Sjöstad is a new part of inner Stockholm. An old dockland and industrial area is being transformed into a new city district with 9,000 flats housing a population of 20,000. The area will have an inner-city character with a mixture of flats, shops, offices and small trade (Figure 9-1). The number of

© IWA Publishing 2006. *Strategic Planning of Sustainable Urban Water Management* edited by Per-Arne Malmqvist, Gerald Heinicke, Erik Kärrman, Thor Axel Stenström and Gilbert Svensson. ISBN: 1843391058. Published by IWA Publishing, London, UK.

people that will be connected to a new wastewater system is assumed to be 15,000; this number has been used in our studies.

The environmental goals were set high in the planning of Hammarby Sjöstad, for example, in the choice of building materials. The technical infrastructure also reflects high environmental ambitions, including energy production and waste management. Goals for water and wastewater include a 50% reduction in water consumption compared with average newly constructed areas, 95% recycling of the phosphorus present in wastewater, local treatment of all stormwater and a 50% reduction of hazardous substances in the wastewater. Stockholm Water Co. is responsible for water and wastewater in the area. The company produces and delivers drinking water to about one million people in Stockholm and ten neighbouring municipalities; it also treats wastewater from Stockholm and seven neighbouring municipalities.

Figure 9-1. The model city of Hammarby Sjöstad from the waterfront.

A local wastewater treatment plant is planned for Hammarby Sjöstad, with new technology for the recovery of nutrients. Biogas produced in the plant will be used as fuel for vehicles and stoves, and the sludge will be spread on arable land. The design of the new plant is currently under development in a pilot plant. Stormwater will be treated locally and will not add to the load on the WWTP. Water-saving equipment is used throughout in order to lower the water consumption. The Henriksdal wastewater treatment plant is located in the vicinity of the model city. As one of the largest WWTPs in Sweden, it is designed for a load of 370,000 m^3 per day and a BOD$_7$ load of 63 tons O$_2$ per day. The treatment consists of pre-precipitation of phosphorus using iron sulphate, pre-sedimentation, active sludge treatment designed for 70% reduction of nitrogen, post-sedimentation and filtration. Sludge is dewatered and anaerobically digested prior to shipment to the north of Sweden, where it is used in land reconstruction around the Aitik mine in Gällivare.

9.1.1 Water systems studied

The system alternatives considered in the analyses deal mainly with treatment of domestic wastewater and stormwater. The supply of drinking water is the same for all system structures and is therefore not studied specifically.

Drinking water system

Raw water is taken from Lake Mälaren and treated in the drinking water treatment plants of Lovö and Norsborg. The treatment train of the two plants is almost identical and consists of coagulation and sedimentation followed by rapid sand filtration, slow sand filtration, disinfection with UV and/or chlorination, and adjustment of hardness and alkalinity. From the treatment plants, the water is distributed to Hammarby Sjöstad by means of the existent distribution system. Within the model city, a new distribution system is constructed without using materials that may have negative effects on health and environment, such as copper and PVC. The water consumption is assumed to be similar for the system structures discussed, and it is assumed that flats are equipped with water-saving equipment, such as low-flushing toilets. The water usage is estimated to be approximately 140 l/(p*d)[1]. The first measurements within the area indicate that the water usage is slightly lower than the average usage in Stockholm (Gannholm 2004; Magnusson 2003).

System structure 1: Combined system with centralised treatment of stormwater and wastewater

All wastewater and polluted stormwater in system structure 1 is collected by a combined sewer system and led to the central wastewater treatment plant (Figure 9-2). Relatively unpolluted stormwater is discharged into the local recipient. The treatment of wastewater consists of mechanical, biological and chemical stages, and includes nitrogen removal. Sludge is treated by anaerobic digestion and supercritical wet oxidation with subsequent sludge fractionation for the recovery of phosphorus. The phosphorus product obtained during fractionation can be

---

[1] In international comparison, the domestic water consumption in Sweden is relatively high.

used as raw material in the production of chemical fertilisers, or directly applied to arable land. Organic waste is collected together with other household waste and transported to an existent incineration plant in the southern part of Stockholm; the ashes are landfilled.

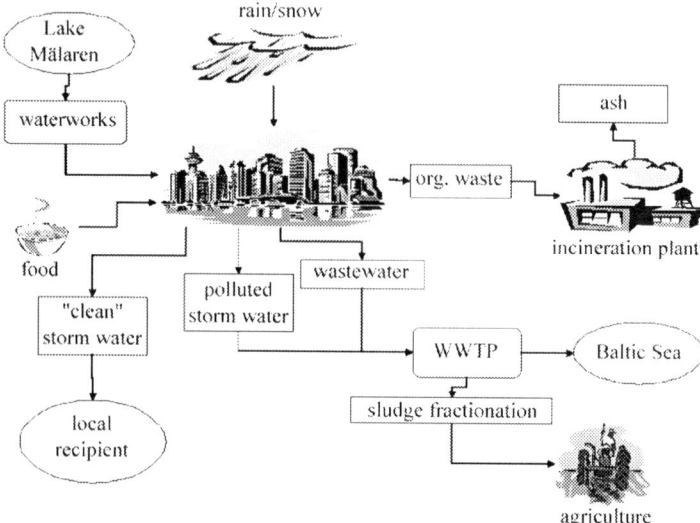

Figure 9-2. System structure 1: Combined system.

System structure 2A: Blackwater and urine diverting system with food waste disposers

This system is based on source separation of stormwater, greywater, blackwater and urine. Stormwater is treated locally, greywater in a conventional treatment plant. Urine and blackwater are separated by means of urine-diverting toilets. The urine is collected and used as fertiliser. Blackwater is digested in anaerobic reactors together with organic waste from households and restaurants, which is added to the sewer by food waste disposers. The system is illustrated in Figure 9-3. Urine is transported in separate pressure pipes to short-term storage in the vicinity. After storage, the urine is transported by truck to farmland for use as fertiliser. Blackwater and organic waste from food waste disposers are transported in separate pipes to a local treatment plant. The treatment consists of pre-precipitation and pre-sedimentation with metal salts for precipitation of phosphorus. Sludge from this process is sanitised, digested and dewatered prior to transportation to arable land. After sedimentation, the blackwater is treated in anaerobic reactors (UASB reactors[1]) for methane gas production. After anaerobic digestion, additional treatment is required. The anaerobically treated blackwater is mixed with greywater and further treated by a conventional activated sludge process, either locally or at the central treatment plant. The sludge is digested, dewatered and transported to an existing incineration plant. The ashes are landfilled. Polluted stormwater is transported in separate pipes to special treatment facilities: sedimentation tanks, filter cassettes and sand filters. The sludge is landfilled, while treated stormwater is led to the local recipient. Unpolluted streams of stormwater are piped directly to the recipient.

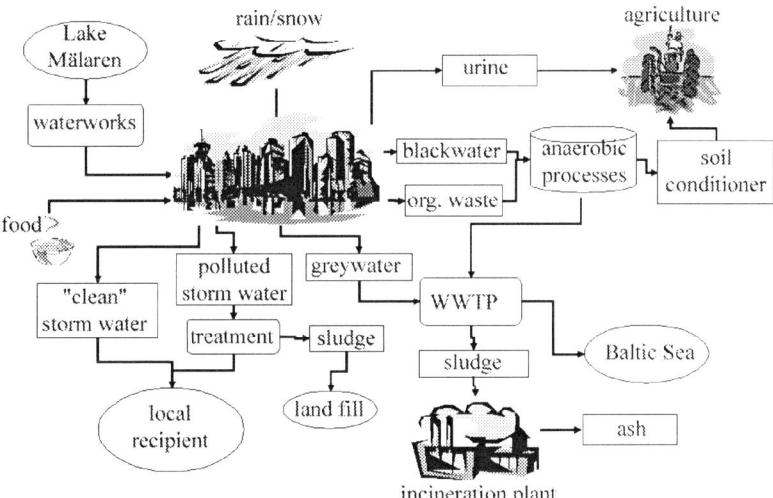

Figure 9-3. System structure 2A: Urine-diverting blackwater system with food waste disposers.

System structure 2B: Blackwater system with food waste disposers and reverse osmosis

System structure 2B is similar to 2A, but does not apply source-separation of urine. The blackwater (which here includes the urine) is treated in a UASB reactor, and concentrated by reverse osmosis (RO). The concentrate is spread on arable land (Figure 9-4).

---

[1] UASB = Upflow Anaerobic Sludge Blanket

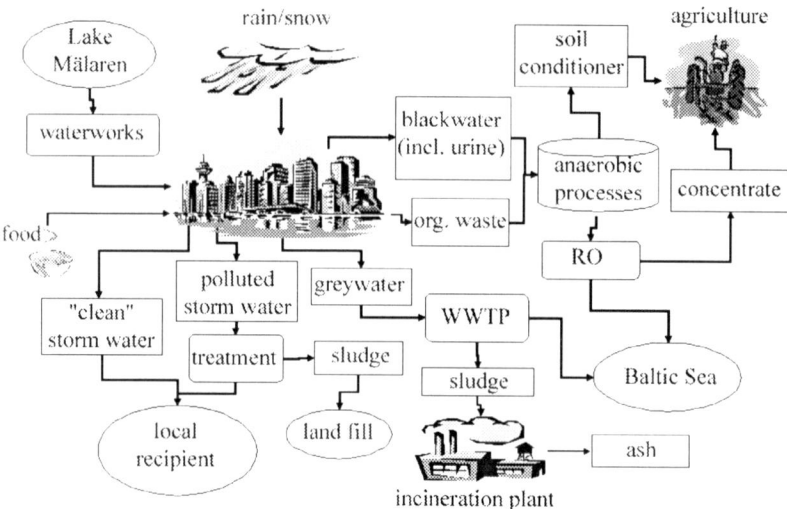

Figure 9-4. System structure 2B: Blackwater system with food waste disposers and reverse osmosis.

System structure 3: Local treatment with nutrient recovery

System structure 3 is intended to represent the system planned for Hammarby Sjöstad as closely as possible. Domestic wastewater and stormwater are kept apart and treated separately. The wastewater is transported through separate pipes to a new local treatment plant. The treatment process consists of mechanical treatment, pre-sedimentation, active sludge treatment without nitrogen removal, filtration and reverse osmosis (RO). The concentrate from the RO and the digested and dewatered sludge are used as fertiliser on arable land. The stormwater is treated separately as in system structure 2. Organic waste is collected separately and transported by lorry to a new digestion facility. The residual product from digestion is transported to farms for use in agriculture (Figure 9-5).

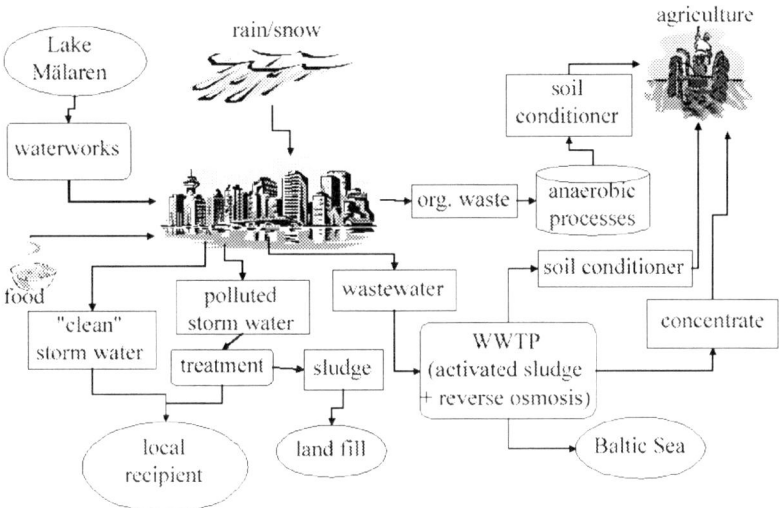

Figure 9-5. System structure 3: Local wastewater treatment with nutrient recovery.
AS = activated sludge; RO = reverse osmosis

9.1.2 System analysis

The system analysis of Hammarby Sjöstad is based on results from the projects presented in Boxes 1 and 2, as well as common knowledge and results from the other Urban Water model cities.

Hygiene and microbial risks

Risk assessments have been made for the combined system structure 1 and the source-separating system structures 2A and 2B (Ashbolt *et al.* 2004; 2005). It is reasonable to assume that the hygienic risks are small for all four systems; thus all of the investigated systems investigated here are acceptable from a hygienic viewpoint. The pathogen group of greatest concern varied by point of exposure; however based on overall risk, enteric viruses generally dictated the outcome in both the centralised and source-separated alternatives at Hammarby Sjöstad. The highest risk of infection resulted from obvious risks, such as exposure to raw sewage during mains breaks in a centralised system structure, or manually unblocking urine-diversion pipes in a source-separating system; yet few individuals would be exposed in such situations and overall community risks were ranked minor for both systems. In the combined system, the most notable risk was the exposure of recreational swimmers. See also Section 3.1.

Specific Urban Water projects in the model city Hammarby Sjöstad

Microbial risk assessment (MRA): The aim of this project was to compare the risk of exposure to pathogens for the alternative system structures using the Microbial Risk Assessment tool, see Section 3.1. The result is primarily a relative comparison of the risk of infection for the system alternatives (Ashbolt *et al.* 2004; 2005).

Chemical risk assessment of wastewater fractions (CRA): Chemical risk assessment was made with the CHIAT tool, see Section 3.2 (Ledin *et al.* 2005).

URWARE: Aspects of environmental sustainability were modelled with the URWARE tool. Included were effects on climate, acidification and eutrophication, exergy consumption and nutrient recovery potential, as well as the quality of residual products (Hellström *et al.* 2005; Jeppsson *et al.* 2005; Jönsson *et al.* 2005a)

Barriers for stormwater systems: The project aimed to identify possible barriers for the prevention of unwanted substances in stormwater, focusing on process and structural barriers. Four types of barriers were identified: organisation, legislation, physical structures and treatment processes, and behaviour related barriers (Karlsson *et al.* 2004, Palmquist 2004).

Comparison of costs for alternative wastewater systems: The costs of proposed wastewater systems in Hammarby Sjöstad were compared, including their distribution between public and private stakeholders. The results of the project illustrate the difficulties in comparing costs, especially when an existent, established system is compared with alternative systems (Olin *et al.* 2005). See also Section 4.1.

Institutional capacity: The objective was to devise criteria for guidance in the strategic planning of sustainable water management systems. In Hammarby Sjöstad, the reasons for not carrying out the plans for a urine diverting system were investigated. (Storbjörk and Söderberg 2003).

Assessment of communication strategies: The communication between the stakeholders in the usage and maintenance of the technical infrastructure was studied to develop an effective cooperative strategy (Drangert *et al.* 2005).

Technical risk assessment was made for the combined systems and for systems based on source separation (Olofsson, B. *et al.* 2001).

Synthesis workshop: To evaluate and further refine a method for synthesis of the system analyses, a workshop was held (Etnier and Söderberg 2002; Söderberg 2002; Söderberg *et al.* 2002)

Box 9-1. Projects in the model city Hammarby Sjöstad, part 1.

PhD projects in the model city Hammarby Sjöstad

Comparative analysis of faecal contamination: Occurrence and reduction of pathogens (enterovirus, norovirus, *Giardia, Cryptosporidium*) in the pilot plant were measured for one year. The data were compared with measurements of common faecal indicator organisms (Ottoson 2005).

Distribution of hazardous substances in the reuse of wastewater nutrients: The aim was to analyse the flow of hazardous substances through society and the wastewater systems. Heavy metals and approximately 80 selected organic compounds were analysed in fractions of household wastewater (Palmquist 2004a; 2004b).

Product recovery by selective sludge fractionation: Techniques for the recovery of phosphorous from sewage sludge, such as leaching with acid or base, were studied (Stark 2005).

Innovative information systems: Control strategies and methods for management of data from a pilot plant were evaluated in order to make the operation, surveillance and control of the treatment plant more effective (Ekman 2005).

Related projects conducted by the Stockholm Water Co.

Blackwater system: The investigations aimed to evaluate the prerequisites for a blackwater system. Relevant criteria were nutrient discharge, nutrient recovery, water usage, exergy-efficiency, economy, operation and maintenance, user aspects, and farmers' attitudes to the fertiliser products (Hellström *et al.* 2003; Hellström *et al.* 2004b; Rydhagen 2003).

Local wastewater treatment: A pilot plant at Henriksdal WWTP evaluates treatment processes for the local WWTP in five parallel lines: (1) aerobic treatment with biological removal of nitrogen and phosphorous; (2) aerobic treatment with membrane bioreactor and reverse osmosis; (3) anaerobic treatment with fluidised bed and reverse osmosis; (4) anaerobic treatment in an Upflow Anaerobic Sludge Blanket (UASB) reactor with biological nitrogen reduction; and (5) anaerobic treatment with membrane bioreactor and RO (Björlenius 2003).

Measuring station: To collect data on the composition of wastewater in the model city, in particular organic substances (Magnusson 2003).

Information technology: Development of real-time information systems for users is planned, with the objective to increase the commitment of the residents to environmental issues related to the transport and treatment of wastewater.

Information centre: In the environmental centre of Hammarby Sjöstad, the technical solutions for waste, energy, water and wastewater are presented in an attractive form, and the responsibility of each resident is emphasized (www.hammarbysjostad.se).

Individual water meters were installed in 161 flats in the area and came into function in January 2005 (Gannholm 2004).

Box 9-2. Projects in the model city Hammarby Sjöstad, part 2.

Environmental effects and resource consumption

The environmental effects were analysed with the URWARE tool (see Section 2.1 and Hellström *et al.* 2004b; Hellström 2005; Jeppsson *et al.* 2005; Jönsson *et al.* 2005a; 2005b). Of the studied categories of environmental effects, *i.e.* effects on climate, acidification and eutrophication, eutrophication was the most important. From this aspect, the local WWTP (system 3) was the most advantageous. The source-separating systems (2A and 2B) also showed low emissions of eutrophying substances. The combined system (1) had twice the emissions generated by the other systems, however they were still in the range of what is considered ecologically sustainable.

Although the local WWTP (system 3) has the highest nutrient recovery potential, it also has the highest exergy consumption. The urine-diverting blackwater system (2A) stands out as an attractive alternative, since it has by far the best net balance in exergy consumption and allows for the recovery of fairly pure nutrients. However, the disadvantage of system 2A is the relatively low recovery potential for phosphorus. If a high recovery level of a relatively clean phosphorus product is required, system 1 is preferable. From a nutrient recovery perspective, an optimised system could be a combination of systems 1 and 2B, *i.e.* a urine-diverting system combined with a sludge fractionation process. If all resource aspects are considered, the blackwater system with RO (2B) seems to be the best solution. This alternative consumes less process chemicals than the recovery of phosphorus by sludge fractionation in system 1 (Hellström 2005).

Chemical risks

A chemical hazard identification and assessment tool (CHIAT, see Section 3.2) analysis has not been specifically carried out in Hammarby Sjöstad due to lack of data for the substances in question, as well as lack of information on treatment efficiencies of the suggested technologies. The discussion of chemical risks presented below is simply a starting point for further investigation.

Generally, the methods for sludge treatment should be optimised to make sure that substances present in the sludge fraction are degraded to a level below the lowest effect concentration, before they are used on arable land. Earlier investigations suggest that, in stormwater, this is relevant for 58 substances. The corresponding number for greywater is 46, for urine 99, for faeces 99 and for organic waste 8 substances (Ledin *et al.* 2005).

For the combined system (1), a potential source of chemical risks is the transport of presumably clean stormwater directly into the local recipient. It is important to investigate whether any of the pesticides (in total 58) that have been classed as potentially hazardous (Ledin *et al.* 2005) are used within the area. It is also necessary to determine whether any of the substances present in the water phase, and identified as potentially hazardous in stormwater (in total 72), greywater (24), urine and faeces (81), will pass through the treatment plant and still be present at concentrations in the recipient at concentrations higher than the lowest effect concentration.

For the urine-diverting blackwater system (2A), chemical risks related to stormwater and the spreading of urine should be evaluated. Greywater, blackwater and organic waste are treated in a treatment plant, *i.e.* the same considerations apply as in the combined system. In the blackwater system with RO (2B), urine is mixed with faeces and organic waste. The same chemical risks apply as for system 2A, with the exception that the treatment efficiency of the reverse osmosis process should be included in the evaluation.

For system 3 (local WWTP), greywater, urine and faeces are co-treated locally. Sludge from the treatment process needs to be assessed for hazardous substances present in greywater, urine and faeces. Also here, the effluent quality should be investigated for possible chemical risks.

Economy

The total yearly cost per inhabitant in Hammarby Sjöstad varies from approximately €125 for system 1 to approximately €260 for system 2A (Hellström 2005). There is only a small difference between yearly costs for the local WWTP and the two source-separating alternatives. The difference between the combined system and the alternative systems is so great that any possible income potential of the increased recovery of nutrients and energy for the alternatives cannot change the economic conclusions. In Hammarby Sjöstad, a newly constructed district in an urban community with an existent wastewater structure, it is not economically viable to invest in source-separating or local systems. It is noteworthy that the difference in costs between the local WWTP (system 3) and the source-separating systems (2A and 2B) is relatively small (15% or less). Earlier studies, *e.g.* Carlsson Reich (2002), have shown the source-separating alternatives to involve considerable additional costs compared with conventional technology. It is interesting to note that the results from this study also indicate that source-separating systems not necessarily imply higher costs than conventional systems for new areas, provided the same demands are made on nutrient recovery, and there is no possible connection to the existing piping system.

Social aspects

The residents of Hammarby Sjöstad have not primarily chosen the area for its environmental profile; they prefer environmental solutions that are part of the installations and that do not require a change in behaviour (Drangert *et al.*

2005). As the residents stated, wastewater is invisible, in contrast to solid waste. This means that the households tend to take less responsibility for the residual products, since they expect others to manage it, *i.e.* the wastewater utility and the building proprietor. On the other hand, the participation of the residents is crucial for the fulfilment of the environmental goals. It is possible to incorporate into the buildings technical solutions that lead to environmentally sound action, and thus reduce the total environmental impact from the area. However, it is also necessary that residents abstain from using hazardous substances in the households, and from placing waste products in the wrong flow. The residents of Hammarby Sjöstad want to have clear and concrete information on how to act in an environmentally sound manner. Experience shows that a considerable and continuous information effort is required from responsible authorities and companies, to disseminate such information successfully.

Organisational and institutional criteria

From the experience in the model city, some conclusions on organisational aspects can be drawn. The combined system (1) means there is no change for the users, the waste management sectors or the proprietors in comparison with a conventional system. Sludge fractionation is a new technology, which means it would have to be approved by the board of the Stockholm Water Co.; municipal city planners would also be involved with regard to localisation. The vision of the Swedish government is to reuse nutrients from sewage sludge in agriculture, which would favour further development of sludge fractionation techniques.

The blackwater systems (2A and 2B) require more involvement of the waste management sector. A good dialogue with future proprietors, farmers and the food industry is necessary. Physical planning is required for the new facilities, and funding to cover the investments. These systems comply with Swedish legislation, if only the planned treatment method produces a product that is clean enough to be spread on arable land. Regarding food waste disposers, a forthcoming European Union compost directive, which may require organic waste to be composted, may prohibit or restrict their installation.

The local WWTP (system 3) is similar to the combined system (1) from an organisational viewpoint. For the future proprietors it demands less investment than source-separating alternatives, and it does not imply behavioural changes for the users. Scientific knowledge of the system alternatives is continuously increasing. For several of the technical solutions included in the system structures, there is however a lack of experience with full-scale facilities.

In Hammarby Sjöstad, there are well-established routines for informing the residents. The responsible organisations are aware that the different system alternatives require specific communication strategies. One problem is that there is no forum where the stakeholders would meet naturally to discuss possible technical solutions.

Technical function and technical risks

The preliminary identification of risks shows that the number of risk events is significantly higher for the source-separating systems (2A and 2B) than for the combined system (1). The combined system (1) has largely the same risk events as a conventional wastewater system. In large-scale systems however, certain risk events affect a larger number of people than in local systems (Olofsson, B. *et al.* 2001).

9.1.3 Syntheses of the results

A synthesis workshop was held in 2001 with representatives of local and regional stakeholders, the Swedish EPA, the Association of Swedish Farmers, consultants, the scientific community, municipal wastewater utilities and the Urban Water researchers. The interpretation of the results was complicated by the preliminary state of the available information. The workshop nevertheless gave valuable input to the programme and the continued development of synthesis strategies. In summary, it can be said that the implementation of source-separating, local solutions is hindered by a lack of experience with similar facilities, lack of common values regarding central issues, and high investment costs.

9.2 A SMALL TOWN: SURAHAMMAR

9.2.1 Description of the area and the systems studied

The municipality of Surahammar, population 10,200 inhabitants, is situated in the middle of Sweden, in the downstream parts of the Kolbäcksån River catchment area. Wastewater from most of the households is treated in the Haga wastewater treatment plant (WWTP). Around half of the households are equipped with food waste disposers that are connected to the sewer system. The food waste disposers are convenient, and are also a way to make use of an over-capacity in the digestion unit at the WWTP.

There is only a short distance between urban and agricultural areas in Surahammar. Consequently, the recycling of nutrients from wastewater to agriculture, which appears feasible, is the main goal of planning ambitions. For example a renovation project plans to introduce separate piping systems for blackwater and organic household waste for a building with 20 apartments, in an attempt to separate clean flows from polluted ones. The nutrient-rich blackwater will be digested in a separate process line at the WWTP, and the remaining biosolids are intended to be used in agriculture.

The scope of the research was limited to the management of wastewater and solid organic household waste, for which three alternative system structures were selected. They represent A) a conventional wastewater and solid waste handling strategy, B) the existing system in Surahammar, and C) a source-separating system to enable nutrient recovery.

In Alternative A, all wastewater is transferred through the sewer to the WWTP. The treatment comprises mechanical, biological, and chemical processes, before the effluent is discharged to the receiving water. The sewage sludge is digested anaerobically, and transported to arable land to be used as a fertiliser. Solid organic waste is collected separately and transported to a composting plant for soil production (Figure 9-6). The soil is used as a construction material, *e.g.* for covering landfills.

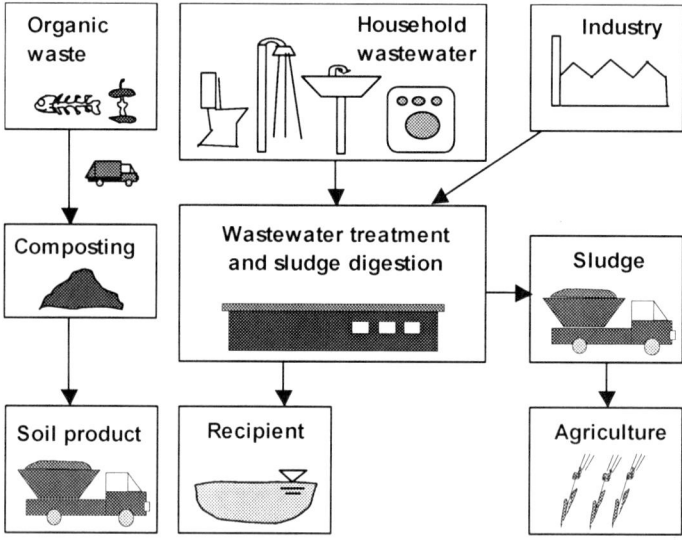

Figure 9-6. A conventional waste handling system (Alternative A).

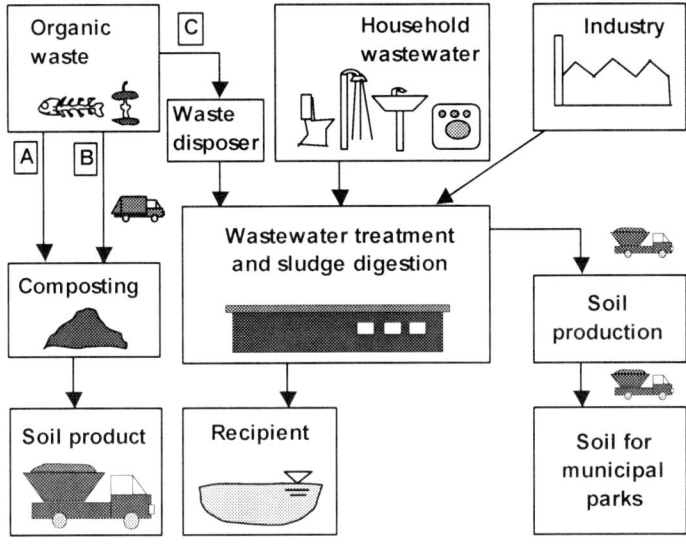

Figure 9-7. The existing waste handling system in Surahammar (Alt. B).

In Alternative B, the current way of handling wastewater and organic waste in Surahammar, there are three alternative systems for organic waste. The majority of the households have chosen food waste disposers, from which the food waste, together with the wastewater, is transferred by the sewer to the WWTP. The two other methods for handling organic waste are to collect it for central composting (as in Alt. A), and home composting. The treatment plant is operated as in Alt. A. Sewage sludge is digested and then transported to a soil-production facility (Figure 9-7).

Alternative C has been formulated to generate a nutrient-rich product, suitable for recycling to agriculture, which originates only from urine, faeces and food waste (not mixed with other waste or wastewater flows). Here, toilet waste from low-flush toilets and milled organic waste are collected in tanks and transported by truck to a separate digestion process; the sludge, transported to arable land, is not dewatered. Wastewater from baths, washing up and laundry (greywater) is transferred by the sewer to a conventional treatment plant; the sludge from the WWTP is incinerated (Figure 9-8).

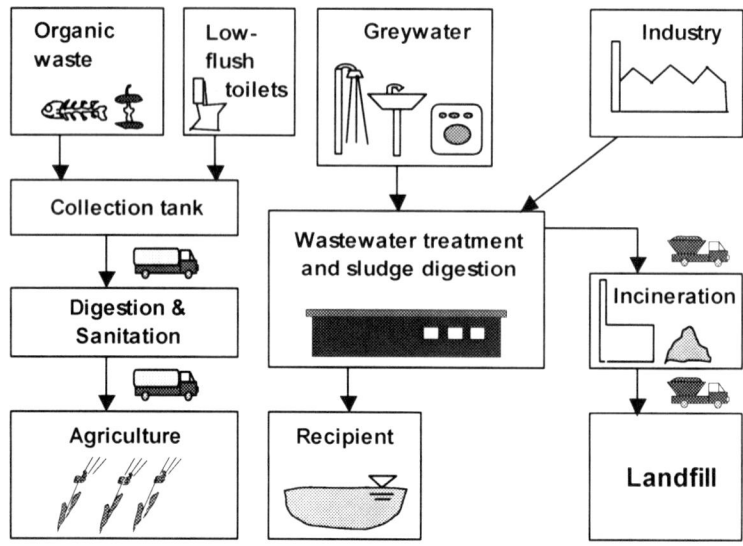

Figure 9-8. Blackwater system for nutrient recovery (Alt. C).

9.2.2 Investigations and systems analysis

The aim of the project was to analyse the sustainability of the alternatives described above. The project group consisted of local actors and researchers from the Urban Water programme. From 2001 to 2003, they met twice a year. The aim of these meetings was to define the scope, select the systems to be studied, and to introduce and follow up ongoing investigations.

Both the environmental assessment tool URWARE (Section 2.1) and the costing tool (Section 4.1) were tested for the first time in Surahammar. A multi-criteria decision aid (MCDA, Section 8) was tested in a hypothetical planning situation, in which real stakeholders together assessed the alternatives presented in Figures 1 to 3 on the basis of an internal report. Researchers from the MCDA project facilitated the process. The NAIADE software was used, which is based on pair-wise comparison of alternatives, followed by an aggregation of all criteria included, and the evaluation of alternatives. The model can handle both quantitative values and descriptions in words.

The internal report included four types of sustainability according to the Urban Water Framework: hygiene, environment, economy, as well as organisation and household. The fifth sustainability aspect in the framework, technical function, was not explicitly included, but influenced indirectly the choice of system alternatives early in the project.

The MCDA process

A multi-criteria decision aid process was introduced in the Surahammar project. A decision-making group (DMG) was formed that consisted of three employees of the municipal service company, an employee from the municipal environmental department, and a farmer. Employees from other departments of the municipality, as well as a representative from an NGO, were also invited, but did not participate. The process consisted of the following steps:

1. *Shaping and designing modes*: Selection of the sustainability criteria,
2. *Comparing mode*: Analysis of the system alternatives, by using Urban Water's tools, and
3. *Choosing mode*: Assessment and comparison of the system alternatives, with the NAIADE tool.

The DMG met three times, and worked through Steps 1 and 3. In the meantime, the researchers prepared the comparison of systems alternatives (Step 2). At present, the project is in the phase of *comparing options*.

9.2.3 Results of the system comparisons

Shaping and designing mode: Selection of criteria

The first task for the decision-making group (DMG) was to select criteria. A list of 13 suggested criteria was presented by the facilitator, from which the DMG was allowed to discard some, or add own criteria, as long as there was consensus in the group. The DMG decided to keep all suggested criteria, but also to add two new ones: transition costs and chemical risk. The list of sustainability criteria used in Surahammar is given below. During the first meeting, the DMG was also introduced to NAIADE.

*   Exposure to pathogens
*   Chemical risk
*   Toxic compounds to water
*   Cadmium to arable land
*   Nutrient recycling
*   Energy use
*   Eutrophying discharges
*   Total annual costs for new construction, incl. investments, operation and maintenance

- Annual transition costs incl. investments, operation and maintenance
- Financial risks
- Opportunity to participate and learn
- Robustness
- Convenience
- Sphere of action, *i.e.* legislative and political support
- Explicit division of responsibilities and risks among the actors

Comparing mode: Analysis of the system alternatives

By using a simplified microbial risk assessment, the criterion *Exposure to pathogens* was analysed (van Moeffaert 2003b). It was shown that bathing in the recipient was the most critical exposure point in alternatives A and B. Another important point of exposure was the handling of sewage sludge during transport and spreading in alternative A. Alternative C, the blackwater system, has benefits compared to the other systems, since the toilet water is collected separately and therefore does not cause discharges of pathogens to the recipient. On the other hand, the more concentrated blackwater needs to be collected, transported and treated. These activities pose a high risk of exposure to the operation and maintenance staff.

For the second criterion, *Chemical risk*, pharmaceuticals were in focus. The faith of pharmaceuticals was assessed based on information available on their behaviour in the environment. It was concluded that alternative C, the blackwater system, had advantages compared to the other systems, due to the application of urine and faeces on soil. In systems A and B, all pharmaceuticals are transported to the wastewater treatment plant, which is an incomplete barrier against pharmaceutical flows to the recipient. The faith of pharmaceuticals in water was considered more critical than their application to soil. More details about the study are given in Kärrman *et al.* (2005a).

The environmental criteria were analysed with URWARE (Section 2.1). The results show that the blackwater system is beneficial for the protection of the recipient from eutrophying discharges and cadmium, *i.e.* the parameter that represents the criterion toxic compounds to water. Nutrient recycling was represented by the indicator phosphorus recycling. For this aspect, the conventional waste handling system (Alt. A) was the most beneficial one, because a high proportion of sewage sludge was used in agriculture. The energy use was similar in alternatives A and B, but almost four times higher in Alternative C. This was due to dilution of the blackwater fraction with flush water, which consumes fossil fuels for transport, as well as electricity for treatment. More details about the URWARE analysis are given in Kärrman *et al.* (2005a). The URWARE analysis was further used in a spin-off project, where a life cycle assessment of the three system alternatives investigated included farming in the system boundaries (see the end of this section).

Three indicators represented the economic dimension. The indicator of Annual costs for new construction, consisting of investment costs, operation and maintenance, showed similar costs for alternatives A and B, but 25 % higher annual costs for Alternative C. The same was found for the criterion *annual transition costs*. Both these criteria were evaluated with the costing tool (Section 4.1). The third economical criterion, financial risks, was qualitatively assessed; considered were the current financial situation, and the present knowledge level. Alternative A and B are both well documented and affordable within the present tariff level. With the blackwater system, there is little experience, which would necessitate additional staff at least during the introduction, as well as further financial uncertainties. This would mean increased tariffs, or the need for external funding (*e.g.* governmental subsidies).

The socio-cultural criteria consisted of three criteria related to households, and two criteria related to organisation and implementation. Critical factors for the households were *Opportunity to participate and learn, Robustness* and *Convenience*. It was concluded that all alternatives were expert driven and none of them demanded efforts from the households. Neither alternative A nor B incurs restrictions to the behaviour from the users, while alternative C means that the households must avoid strong chemicals for cleaning toilets, and also to avoid pouring greywater from scrubbing floors into the toilets. The blackwater system is therefore less robust to misuse than the other two systems. The food waste disposers were regarded more convenient than the collection of solid organic waste. A conventional WC was further considered more convenient than a low-flush toilet used in the blackwater system. Since alternative B has both conventional WCs and food waste disposers, this alternative is considered as the most convenient one. More details about the household aspects are given in Kärrman *et al.* (2005a).

The organisational criteria were assessed in a qualitative way, by analysing laws and restrictions, and interviewing employees at authorities and at the company that operates the water and wastewater systems (Storbjörk and Söderberg 2003). It was concluded that the *sphere of action, i.e.* political and legislative support, exists for all systems. There could be legal obstacles for the management of blackwater, since this is a waste material, the handling of which is regulated by law. Food waste disposers have support in Swedish law, but a future EU-directive for organic waste management proposes a total ban against transporting grinded food waste to wastewater treatment plants, since this increases the volume of sewage sludge. Regarding the *Explicit division of responsibilities and risks* among the actors, it was concluded that all systems were equal, since the municipal service company would be responsible in any of them. The blackwater system however would affect more actors than the other two alternatives.

From the total number of 15 criteria, eight criteria were given qualitative scores, and 7 quantitative ones. The qualitative scores were given on a 9-point scale from *Excellent* to *Very bad*. For the quantitative scores, NAIADE did not limit the scale.

Choosing mode: Assessment and comparisons

The next task for the decision-making group (DMG) was to assess the criteria one by one. The procedure for assessing the qualitative scores was to accept or adjust the expressions proposed by the authors of the internal report. The revised qualitative scores were placed in a performance matrix. An example of the final score of the qualitative criterion convenience and the quantitative criterion annual costs for new construction is given in Table 9-1. The quantitative criteria were placed directly in the performance matrix.

Table 9-1. The evaluation of the three system alternatives in terms of the criteria convenience and annual costs, which includes investments and operation & maintenance costs.

	A. Conventional	B. Existing	C. Blackwater
Convenience	Neither good nor bad	Good	More or less good
Total annual costs for new building M€/yr	2.0	2.0	2.4

The group should then agree on limits where a difference between two scores should be considered as *much greater than, greater than, approximately equal to, very much equal to, less than* and *much less than*. The NAIADE model uses an indirect way of weighting that prioritises those criteria with scores much greater than, or much less than. The result of this scaling process was that the highest indirect weight was given to the following parameters:

- Total annual costs for new building,
- Annual transition costs,
- Nutrient recycling, and
- Eutrophying discharges.

NAIADE ranked the alternatives: A as the most preferable, followed by B or C, in indeterminate order. The reason behind this ranking can be summarised: Alternative A has the beneficial combination of moderate costs and a high degree of phosphorus recycling, while alternative B has moderate costs but lacks recycling of nutrients. Alternative C recycles the nutrients and causes the least eutrophying discharges, but is ranked low because of the high costs.

Before the NAIADE results were presented, each member of the DMG was asked to make his or her own overall assessment of each system alternative. Even though the participants agreed on the assessment of each criterion, their overall assessment of the system alternatives alternative was disparate.

9.2.4 Lessons learnt

Experience from testing NAIADE in Surahammar and other case studies show that the structured procedure for discussion of the criteria one by one was considered to be advantageous, since a decision-making process with as many as 15 criteria needs supportive tools. The participants found that the process with NAIADE support was very instructive. The researchers concluded that the NAIADE model was too complex for this case study. As a result, the users showed too much respect for the model and were uncritical of the outputs. For strategic decision support, a simpler and more transparent model, like the Strategic Advisor (STRAD), could be more advantageous (See also the Uppsala case, Section 9.5).

9.2.5 Other Surahammar projects

As a spin-off from the investigations described above, more detailed research was conducted in Surahammar. The short summaries below may provide readers with further ideas for their own applications.

An assessment of hazardous flows

An approach to a chemical risk assessment tool with a barrier was developed by Palmquist (2004a; 2004b), who also carried out a comparative substance flow analysis of two wastewater management scenarios for Surahammar, in which the conventional system (Alt. A) was compared with the source-separating one (Alt. C). The study included 16 representative hazardous substances, *i.e.* their presence in grey- and blackwater, and their reduction rates in the WWTP. A possible management approach was suggested: to assess system alternatives and to compare how much the flow of hazardous substances can be reduced, diverged, or transformed, at the source or during transport throughout the system. System design, process barriers, and organisational and behavioural barriers were suggested (Section 3.3). The effect of system design and process barriers was analysed for the fate of phosphorus, cadmium, and triclosan. The results show that the combined system caused a higher substance flow to the receiving water body than the separated system. In the combined system, more phosphorus and cadmium were transported to the farmland than the separated system, but only half the amount of triclosan (Malmqvist and Palmquist 2005).

Wastewater Management Integrated with Farming

The main objective of a study carried out by Tidåker *et al.* (2005) was to analyse the environmental impact and resource use in systems that integrate wastewater management and agriculture. The system alternatives (Figures 9-6, 9-7, and 9-8) were analysed by a combination of LCA with simulations using URWARE. The system boundaries included agricultural activities, such as the production of commercial fertiliser and fuel, transports, soil preparation, sowing, spraying, fertilising and harvesting.

The blackwater system required slightly more primary energy than the other two systems. In particular, the construction of storage facilities contributed considerably to the energy use.

The emissions of greenhouse gases and SO_2 were of the same magnitude for all three systems, while the eutrophying emissions were reduced significantly in the blackwater system. As regards NH_3 and NO_X, the emissions were highest for the blackwater system.

High substitution of mineral fertiliser by recycled nutrients, an optimal spreading technique, and well-designed collection and storage facilities were important factors for the environmental outcome in the blackwater system.

By separating blackwater from the remaining wastewater, a major proportion of the nitrogen and phosphorus found in the sewage was recycled to arable land, and the eutrophying emissions from the wastewater treatment plant were reduced considerably. The decreased emissions of nitrogen and phosphorus by the blackwater system were the most significant difference, compared with other anthropogenic impacts. This indicates that blackwater systems are primarily of interest in locations where the eutrophying emissions are critical.

Sustainable development and urban water management: Linking theory and practice of economic criteria

In his doctoral thesis, Hjerpe (2005) analysed the potential for using economic criteria to assess the sustainability of urban water management. A set of economic criteria was chosen for the maintenance of the water infrastructure, affordability, cost-recovery, effectiveness and potential for development. For each criterion, indicators were chosen and applied to three changes in infrastructure in Swedish municipalities. One of these projected changes was the introduction of kitchen waste disposers in Surahammar. The study found that using economic criteria and indicators in relation to sustainable development requires a continuous balance between the universal and the context-specific, that is, between the criteria and indicators used and the water infrastructure being assessed. This emphasises that the criteria applied should relate to all of the dimensions of sustainable development. It also points out the importance of involving the stakeholders in such assessments.

9.3 A CITY CENTRE: VASASTADEN IN GÖTEBORG

For a part of Göteborg, a systems analysis of ways to handle stormwater and domestic wastewater, which relied mainly on the SEWSYS substance flow model (Section 2.2), was conducted. Four system alternatives were compared: the existing predominantly combined sewer; a separate sewer; a source-controlled separate one; and a combined one complemented with a blackwater system. The aspects considered were substance flows, effects on the environment, the recycling of nutrients contained in wastewater, costs, as well as robustness and risks. Finally, the benefit of source-control measures in storm- and wastewater was simulated. This chapter provides the reader with a brief summary of this comprehensive project. A full report is available in Swedish (Ahlman *et al.* 2004b).

9.3.1 The Vasastaden catchment

The Vasastaden district in the city centre of Göteborg is densely populated, and consists of residential and commercial buildings, built mainly in the late 19th century and the early 20th century. For the case study, Vasastaden has been divided into six subcatchments, according to the existing sewer systems, the character of the buildings and surfaces, and the spatial relation to an existing bedrock tunnel for the transport of domestic wastewater to the treatment plant. Catchments numbered 1, 2, 4 and 5 are areas typical of cities built in the 19th century, while number 3 has residential buildings mostly from the 1970s, and the 6th one contains larger detached houses. The district also includes two parks, 4 and 6 ha in size (Figure 9-9).

The natural recipient for stormwater, Vallgraven, located directly North of the area, is a canal, which cuts through the city centre. The model requires input data that characterises the area investigated (Table 9-2). Spatial data was imported from a GIS database. The catchment has a total area of 75 ha, 51 of which were assumed to be impervious. The local planning office provided information regarding the number of residents, traffic, and the technical infrastructure. The input data was further complemented by a field study of roof materials. For the stormwater module, local statistics for the frequency and intensity of precipitation were used; as were data for the efficiency of the WWTP.

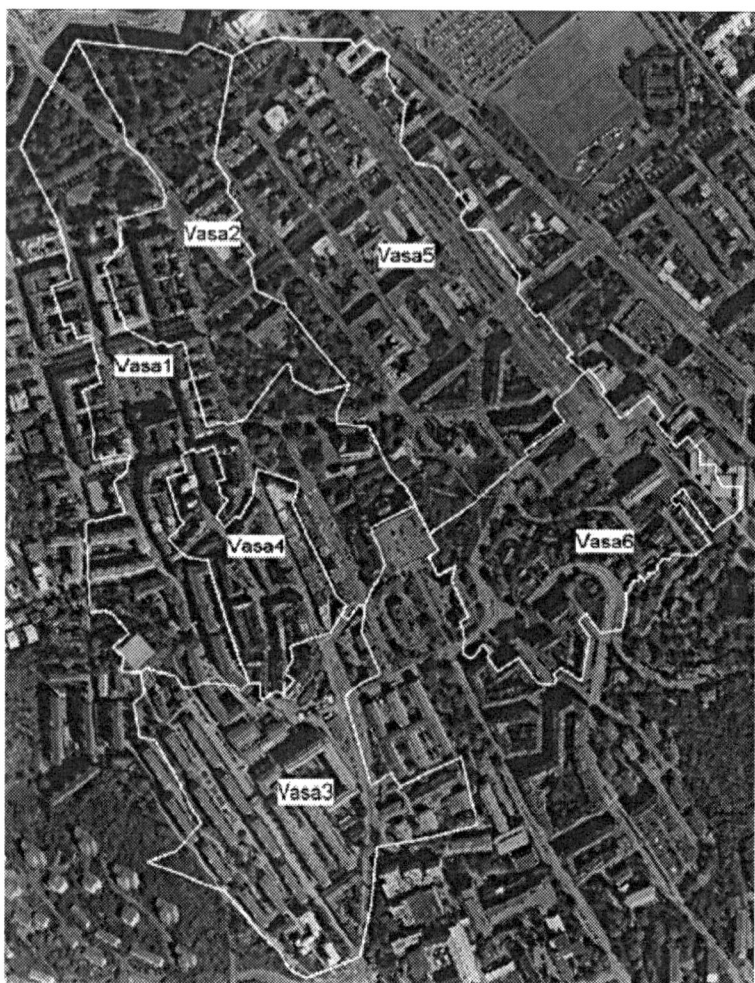

Figure 9-9. Aerial photograph of the six Vasastaden subcatchments.

Table 9-2. Data for the catchment area Vasastaden.

Category	Extent	Unit
Total impervious area	505 600	m²
Roads	223 900	m²
Zinc surfaces near beside roads	2	%
Roofs	204 400	m²
Zinc roofs	60	%
Copper roofs	11	%
Other impervious area	77 300	m²
Total vehicle km	62 900	km/day
Heavy vehicles	5	%
Number of persons connected to the sewer system	8 648	persons
Infiltration water factor	0*	-

* Not considered in this study

9.3.2 System Alternatives

The existing sewer system

There are two kinds of sewer systems in the catchment today: a combined system, and a separate one (Figure 2-11 in Section 2.2). Catchments 4, 5 and 6 have a combined system, in which the sanitary wastewater and all stormwater are conveyed to the wastewater treatment plant (WWTP) in one pipe. Catchments 1, 2 and 3 have a sanitary wastewater system, in which the wastewater is diverted directly to the WWTP, *i.e.* without passing combined sewer overflow (CSO) points. However, since the stormwater is considered too polluted to be diverted directly to the local water body, it is transported to the nearest combined network. Hence, all of the stormwater influences the CSO volumes, and the sewage system of the six catchments as a whole is of a predominantly combined character.

A separate sewer system

In a separate sewer system all stormwater is discharged to the recipient (Vallgraven canal) without any treatment. All the sanitary wastewater is treated at the WWTP; there are no CSOs.

A separate sewer system with source control and treatment

To remedy the problem of direct discharge of polluted stormwater, for a separate sewer system, it is assumed in the model that source control measures and treatment in a retention pond take place before discharge (Figure 2-12). This includes all stormwater from catchments 1, 2, 4 and 5, and that from roads in catchments 3 and 6. The remaining, relatively unpolluted stormwater is infiltrated locally. Provided the state of the groundwater is taken into consideration when stormwater is infiltrated, the ground can be a suitable recipient. Both contaminated soil from the point of infiltration and the sediment from the pond need to be taken into account.

The present system complemented with blackwater handling

To facilitate recycling of a clean nutrient fraction in a sewer system, a separate collection of blackwater may be advantageous. The present sewer system can be complemented with separate blackwater handling in at least two ways (Figure 9-10): one option is to lay additional blackwater pipes to a special treatment facility, the other to store the blackwater locally and transport it by lorry. In this study, only the alternative with pipe transport was investigated further, since a pre-assessment of the costs and the local environmental effects showed that a lorry-based system would be unreasonable.

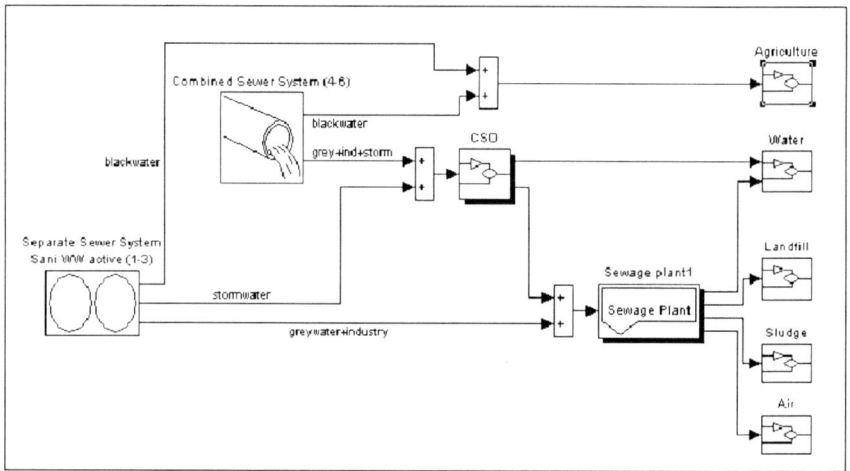

Figure 9-10. Matlab Simulink model for the present Vasastaden system complemented with blackwater handling.

9.3.3 Substance flows in Vasastaden

The total substance flow from the area is shown in Table 9-3. Heavy metals are fairly evenly distributed between domestic wastewater and stormwater, whereas for phosphorus, nitrogen and BOD, the dominant source is domestic wastewater.

Table 9-3. Total substance flows [kg/year] in Vasastaden

	Cu	Zn	Pb	Cd	P-tot	N-tot	PAH	BOD
Sanitary wastewater	29	66	3.1	0.16	6.4	44	0.09	145 000
Stormwater	42	66	3.1	0.10	24	381	0.30	2 200
Total	71	132	6.2	0.26	30	425	0.39	147 000

The model allows for displaying the flows in greater detail. In Figure 2-13, the total substance flow for Vasastaden is divided into sources, in relation to the total load. The largest share of the phosphorus and nitrogen originate from urine and faeces, which represent 50% (N) and 80% (P). Greywater is an important source of heavy metals and BOD, whereas stormwater contributes mainly to heavy metals and PAH[1]. The stormwater may be divided into contributions from wet deposition (pollutants from rain water), from roads and traffic, and from building materials (Figure 2-14). The most polluted areas in Vasastaden are the roads. Figure 9-11 shows that the dominating source of copper in road runoff is break-liner wear. For zinc, lead, cadmium and PAH, tyre and road wear are important sources.

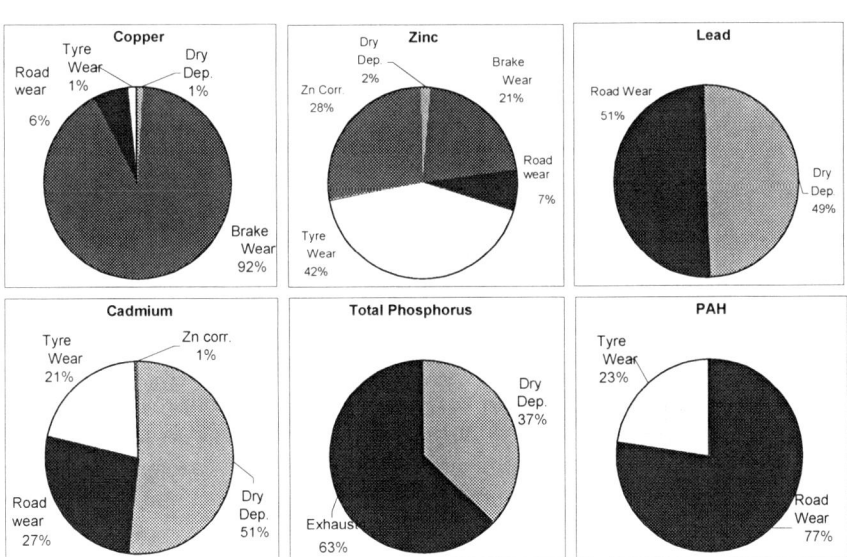

Figure 9-11. Pollutant sources for stormwater from roads.

---

[1] PAH: Polycyclic aromatic hydrocarbons

For the present sewer system, the fate of the substances investigated is shown in Figure 9-12. Since the system is in practice a combined one, the local recipient, the Vallgraven canal, receives pollutants only during CSO events. Except for nitrogen, which may leave the system via nitrification and denitrification, the pollutants are predominantly transported to the sewage sludge. Corresponding calculations were made for all system alternatives.

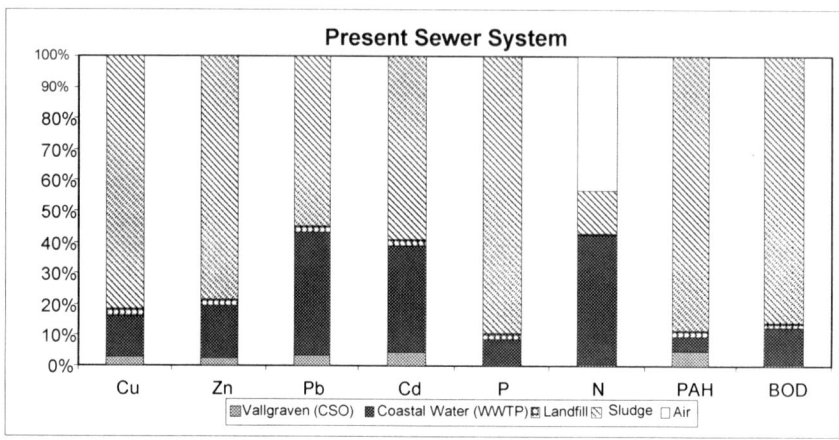

Figure 9-12. Recipients in the present system in Vasastaden.

9.3.4 Environmental aspects

The results of the SEWSYS simulation were used to estimate metal concentrations in the water, in the sediment of the local recipient, and in sewage sludge. To judge whether or not the discharge of a given amount of a substance has a negative effect, the environmental quality standards of the Swedish environmental protection agency (SEPA) were applied for water (SEPA 2000; 2002), sludge (SEPA 2002; 2004), and soil (SEPA 1996; 1999).

The environmental effects of stormwater originate mainly from the heavy metals and PAH, while environmental impacts of sanitary wastewater are due mainly to phosphorus, nitrogen and BOD. It is remarkable that also the upgraded and source-separating sewer systems have a negative effect on almost all of the recipients.

With the present combined system, the CSO volumes are small, so that the nutrients released do not have an environmental impact on the canal. The metals, however, contained mainly in the stormwater, deteriorate the sewage sludge. Cadmium was found to be the most critical metal in sewage sludge.

In the separate system, all stormwater is discharged untreated, and the metals contained in it would add a considerable environmental burden to Vallgraven. Small amounts of metals already negatively affect the reproduction of aquatic organisms; the critical ones in this study are copper, zinc and lead. On the other hand, a separated system would improve the quality of the WWTP sludge. However, this would be an advantage only if there were more catchments with separate sewer systems in the city.

In a source controlled separated stormwater system, less stormwater would reach both Vallgraven and the WWTP, which would improve both the local water body and the sludge quality. However, local treatment requires that the sediment from the retention pond be disposed of, and the transport of pollutants to the groundwater be monitored.

9.3.5 Risks and system robustness

Risk is a function of the probability that an event will occur, and the consequences of it. In this study, the first risk assessment method developed in the Urban Water programme (Olofsson, B. *et al.* 2001) was refined, based on established risk models (Grimvall *et al.* 2003). Risks were assessed as technical, environmental, or health and hygiene.

The probability of a risk event was rated from 1 to 5, when it happens:

1)   Once every 50 years,
2)   Once a year or less frequently,
3)   Once a month,
4)   Every day, or
5)   Continuously.

The consequences were rated correspondingly, here exemplified for the environmental risks in the aquatic environment, the local sediment, and the sludge:

1)   No effect on the environment, according to SEPA standards;
2)   Elevated concentrations, but no effects;
3)   An effect on the environment, according to SEPA standards;
4)   A major effect on the environment, according to SEPA standards; and
5)   All animal life wiped out.

The total risk was calculated by multiplying the probability and consequence for each risk event; the consequences were analysed separately for water, soil and sludge, and then summed over the categories investigated.

Technical risk events taken into account in this study were, for example: clogging of the sewer system, wrongly connected pipes, power failures, low oxygen concentrations with H_2S formation, breakdown of equipment, precipitation in the pipes, incorrect usage, and leaks. The risk analysis indicated that the existing combined system had the lowest technical risks. The two alternative systems, separated with source control, and blackwater, had equally elevated technical risks.

Robustness measures the capacity of a system to resist temporary outer disturbance, and was here estimated in way similar to risk. The robustness is a function of three aspects: the resistance of a system to outer disturbances, its resilience (tendency to return to the former, correct state), and reparability, *i.e.* the time it takes the system (or the operator) to solve the problem when it has occurred. Each of these three factors was rated from 1 to 5, and the robustness calculated by multiplying them for each risk event. The differences between the system alternatives were minor for this parameter.

9.3.6 Economy

In the Vasastaden project, a costing model for the wastewater system was developed. The model takes into account the following types of costs: initial investment, operation and maintenance, and renewal. The model used here differs from Urban Water's costing model (Section 4.1) in that it does not include costs for the existing WWTP; instead it includes the variables "expected interest rate" and "depreciation time". In the model used here, future costs were discounted to present values that would enable comparisons of system alternatives in which costs occur at different times. Some of the most important factors that affect the costs of changing an existing wastewater system are the following:

- The trench depth to lay the pipes,
- The pipe dimension,
- The type of area (here: city centre or not, the type of traffic in the street); and
- The type of ground (here: bedrock or not).

For both the structure of the model and the input, site-specific information was used. The model was built as an Excel spreadsheet. Further details are available in Ahlman *et al.* (2004b). The results make clear that any change of the sewer infrastructure would imply major investment costs. This was true particularly for complementing the present system with additional blackwater handling, the costs of which would become 20 times higher than those for the renewal demand of the present system. The same ranking was found for the operating costs; the blackwater system would be four times more expensive to operate than the present one, and the other system alternatives were between these.

By combining the results from the substance flow modelling and the economic investigation, it was possible to quantify the costs of nutrients recovered in the investigated system alternatives. The costs for nitrogen and phosphorus from sewage sludge, including costs for the change of the system, were about 30 times (N) and 200 times (P) higher in the alternative systems than if nutrients in the present system were used in agriculture. The current cost of N and P from mineral fertiliser is another order of magnitude lower.

9.3.7 Management options

To reach the SEPA standards for the application of sewage sludge to agriculture (*i.e.* the ratio P to Cd), the average cadmium concentration must be reduced by 62%. Also the concentrations of lead and copper are elevated, although not critically. To reduce cadmium flows, measures have to be taken in the whole catchment area connected to the WWTP. Wet and dry depositions, the largest sources of cadmium in Vasastaden, are impossible to prevent by local measures. With the present sewer system, the cadmium concentration in stormwater can be lowered by reducing traffic (tyre and road wear), and by replacing zinc roofs with other materials. The sources of cadmium in the domestic wastewater are diffuse, and mostly unknown. Greywater contains elevated loads; therefore, the content of cadmium in household products should be reduced. If the present sewer system were changed to a separate one, none of the stormwater would reach the WWTP, which would lower the amount of cadmium. Also the introduction of a blackwater system would reduce the flow of Cd to soil, since only source separated blackwater would be used as fertiliser.

If the sewer system were changed to any kind of separate system, measures would have to be taken to reduce stormwater pollution. To reach the environmental standards in the recipient canal, Vallgraven, the concentrations of copper, zinc and lead would need to be reduced by 90%, 55% and 50%, respectively. To reduce copper by 90%, copper roofs would have to be exchanged or painted over, and Vasastaden would have to be closed to car traffic. A large share of the zinc load is due to tyre and break wear, as well as roofs and other zinc-plated surfaces. The greatest source of lead is tyre wear and dry deposition. The following source control measures were simulated.

To alleviate stormwater pollution by metals, technical changes were listed that appear possible to implement by the year 2020:

- To reduce copper fluxes from roofs by 80%, by painting them, and by installing local copper traps;
- To reduce zinc fluxes from roofs and other galvanised surfaces by 20%, by painting or by replacement, which would also decrease the fluxes of cadmium and lead;
- To lower the flux of Cu and Zn from brake linings, by legislation, within 8 years.
- To lower the Cd flux from tyre wear by 80%, by using tyres with 1 ppm instead of today's 5 ppm of Cd;
- To reduce the wear of asphalt from road surfaces by 36%;
- To diminish traffic by 10%; and
- To reduce dry and wet deposition of metals by 1% per year.

The corresponding list for domestic wastewater was:
- A reduced flux of Cd from metal surfaces by 1% per year, an assumption based on the SEPA policy to phase out the use of cadmium in society;
- A decreased Cd flux from household chemicals (by 10%); and
- A decreased flux of Cu, by 20% to the year 2020, achieved by a ban of copper pipes for in-house installations.

These management options were simulated by SEWSYS (Table 9-4). However, this relatively extensive list of source control measures would not achieve the desired reduction of the cadmium flux by 62%.

Table 9-4. The simulated amount of diffuse pollution generated in Vasastaden, at present and in 2020, assuming source control measures. [Kg/year]

	Cu	Zn	Pb	Cd	P	N	PAH	BOD
Present system	71	131	6.3	0.26	6 400	44 200	0.39	147 000
With source control in:								
Stormwater	39	114	5.4	0.24	6 400	44 100	0.24	147 000
Domestic ww	64	128	5.9	0.23	6 400	44 200	0.39	147 400
Reduction by 2020 (in %):								
Stormwater	46	13	14	9	0.1	0.2	39	0.2
Domestic ww	10	2	5	12	0	0	2	0
Total	56	15	19	21	0.1	0.2	40	0.2

9.3.8 Discussion, and concluding remarks

The systems analysis of ways to handle stormwater and domestic wastewater conducted here produced results in several separate fields. Based on the substance flows simulated in SEWSYS, the environmental performance, costs, robustness, and risks were compared for the present wastewater system and selected hypothetical system alternatives. In a decision-making process, these fields need to be weighed against each other, and assigned priorities. However, some general conclusions can be drawn from the data.

If economic aspects alone are taken into account, no changes should be made to the present sewer system in Vasastaden. Both the operating and investment costs of the blackwater system are unreasonably high. The separate stormwater systems (with and without treatment) have only slightly higher operating costs than the present one, but these require additional investments. Since the risk of failure tends to increase with complexity, the present sewer system involves less technical risk than the alternatives.

The interpretation of the environmental analysis is not straightforward. The best solution would probably be a system with separate collection of domestic wastewater and stormwater, with control of pollution sources, and with local treatment of stormwater. However, all of the sewer systems investigated were found to produce sewage sludge which contains heavy metals concentrations that are too high. Considering also the high costs for nitrogen and phosphorus from the sludge, reusing them as fertilizer in agriculture is questionable, at least if this requires major technical changes.

Extensive source control measures are necessary to fulfil current environmental standards for the recipients, sewage sludge, and sediments. Many of the sources of diffuse pollution depend directly on the design of our urban infrastructure (building materials, roads, etc.), and on our lifestyles (traffic, purity of household chemicals). Some of the necessary abatement strategies would counteract ambitions in other sectors of society (mobility, jobs), and are therefore difficult to implement. Reasonable yet effective changes are possible only with a very long time perspective and on a large spatial scale. As long as the sources exist, pollutants will accumulate in one or more of the environmental compartments, be it water, sediment or sludge. Further discussion of the issue of diffuse pollution is given in Sections 3.2 and 3.3.

It can be concluded that substance flow analysis, combined with the investigation of other important aspects, is a suitable tool to shed light on the sources, flows, and fate of substances that constitute the diffuse pollution, and to evaluate hypothetical system changes and management options in a rational and transparent way.

9.4 URBAN ENCLAVES: GEBERS AND VIBYÅSEN

The type of model city termed Urban Enclave, though situated within a city, is characterised by a high level of autonomy in terms of the water and wastewater systems. A special case is an area that is totally independent of the water and wastewater system of the surrounding city. However, the areas presented here, Gebers and Vibyåsen, were not totally independent since they both received their drinking water from the water system of the surrounding urban system. Gebers was studied from many angles including that of comprehensive system analysis, while the studies in Vibyåsen have been limited to measurements on source-separated greywater and blackwater.

9.4.1 Gebers

The housing co-operative Gebers (Figure 9-13) is situated by Lake Drevviken and surrounded by a natural reserve, in spite of its location only 13 km from the centre of Stockholm. It consists of two buildings, the larger of which (2300 m^2 living area) was originally built in 1936 as a convalescent home, and a small one-storey building, the *Blue House*. In 1998, the buildings were re-constructed as an ecological housing co-operative with 30 apartments in the main house and two in the Blue House. In the main house, a large communal kitchen and dining room were left, and here dinners are served four days a week.

Figure 9-13. View of Gebers from the southeast. Photo by Klas Öster.

The grounds are quite large, 3.2 ha, but the soil cover is thin and the bedrock is exposed in several places. The flows and compositions of the wastewater fractions were analysed for the main house in October 2001; at that time, 52 adults (36 women and 16 men, of which one was above 70 years of age) and 27 children (of which 4 used diapers) lived in the house. Gebers receives its drinking water from the Stockholm piping system. The water is produced from Lake Mälaren, by flocculation, sedimentation, rapid and slow sand filtration and disinfection with, depending on treatment plant, either chlorine gas or ultraviolet light combined with sodium hypochlorite.

Gebers has a urine-diverting wastewater system with dry faecal collection (Figure 9-14). Thus, three source-separated fractions are produced: urine, faeces and greywater. The toilets in the main house are located on the 1st, 2nd and 3rd floors and from each a galvanised sheet steel pipe (Ø 200 mm) leads directly down to a collection bin, a plastic 140-litre waste container, in the basement. To control the fire hazard, each bin is placed in a separate sheet-metal closet (Figure 9-15). The closed closets are fan ventilated to decrease the risk of odour in the toilet, and to partially dry the faeces (Andersson and Jensen 2002). The urine is piped from the toilets to three plastic collection tanks in the basement. When these are full, the urine is transported 39 km to a farm, where it is stored at least six months before being used as fertiliser for cereal production.

The two apartments in the Blue House also have urine-diverting dry toilets, but the faeces are collected directly below the toilets. These two toilets have their own urine collection tank. According to the statutes of the Gebers housing co-operative, each household is obliged to manage its own dry toilet system following the instructions of the co-operative. These instructions include an obligation to transport the faeces to the designated faecal compost site on the premises. When the faecal matter has been emptied into the compost, the responsibility for the faeces, its treatment and the compost product is taken over by the co-operative. Bulking agents (leafs and dry grass) are added during the process. The co-operative has an agreement with a farmer, 11 km away, who can use the compost as fertiliser.

However, the volume of the composted faeces after six years is still small enough to be stored in the compost container; therefore no compost has yet been transported to the farm. The greywater is piped to the Stockholm wastewater system and is treated at the large Henriksdal wastewater treatment plant (WWTP). The WWTP effluent is emitted to the Baltic Sea. The stormwater from the roof and the paved area in front of the building is led by a short pipe directly into the local receiving water, Lake Drevviken.

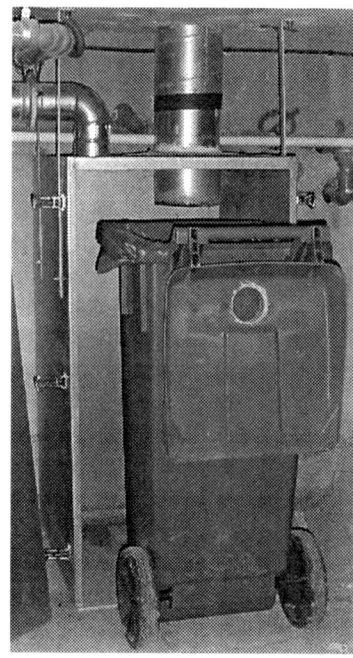

Figure 9-14. Urine-diverting dry toilet system   Figure 9-15. Faecal matter collec-
at Gebers. Drawing: HSB Stockholm.   tion bin. Photo by Annika Jensen.

9.4.2 Vibyåsen

The Vibyåsen housing area consists of terraced and semi-detached houses, situated in the municipality of Sollentuna, approximately 18 km from the centre of Stockholm. It is surrounded by forest and located next to a large nature reserve, Järvafältet. When the greywater was analysed, Vibyåsen comprised 132 households with a total of 381 residents. The drinking water at Vibyåsen is distributed by the municipal network and produced from Lake Mälaren.

The blackwater is source-separated and collected separately from the greywater. Ordinary low-flush toilets are used, and from these the blackwater is piped to a collection tank. At the time of the measurements, the collected blackwater was transported a few kilometres to a nearby farm, where particles were separated. After sanitation by co-composting with animal manure, the separated solid fraction was used as fertiliser. The liquid fraction from the separation was stored in closed tanks for at least six months and then used as fertiliser, in the same way as dilute animal slurry. At present, however, the blackwater is, mainly for economic reasons, piped to the municipal wastewater system.

The greywater is collected in a separate piping system. At the time of the measurements, it was treated by a local treatment facility, consisting of sedimentation and a bio-rotor, followed by sand filtration. The effluent was required to comply with the hygienic requirements for bathing water. After treatment, the water was discharged into a pond system that led to a local forest brook (Figure 9-16). At present, also the greywater is piped, for economic and functional reasons, to the municipal wastewater system.

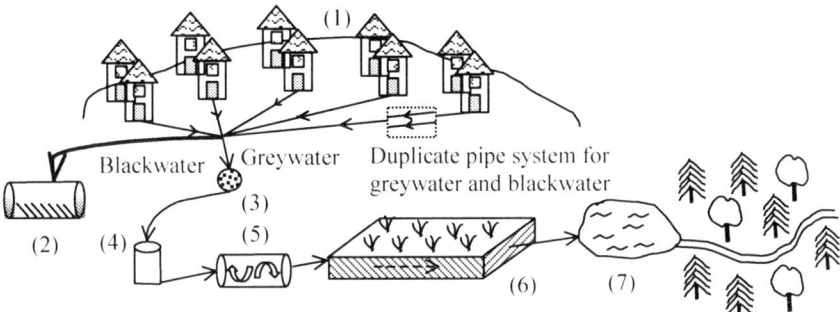

Figure 9-16. Schematic representation of the wastewater system in Vibyåsen: 1) houses, 2) blackwater collection tank, 3) greywater sampling well, 4) septic tank, 5) biorotor, 6) sand filter and 7) pond. Revised from Helena Palmquist.

9.4.3 System structures studied

For Gebers, environmental systems analyses and quantitative microbial risk assessment were made for three system structures: The present system (Figure 9-17), local treatment of greywater (Figure 9-18) and a conventional system as reference (Figure 9-19).

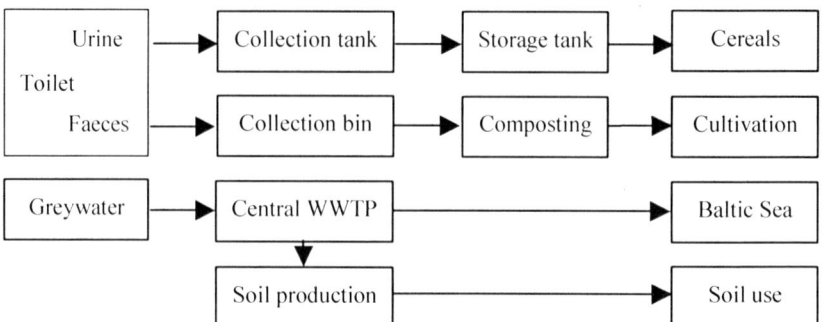

Figure 9-17. Present system: Source-separation of urine, faeces and greywater. Urine transported 39 km for reuse and greywater treated at large municipal plant. Stormwater discharged to the local receiving water.

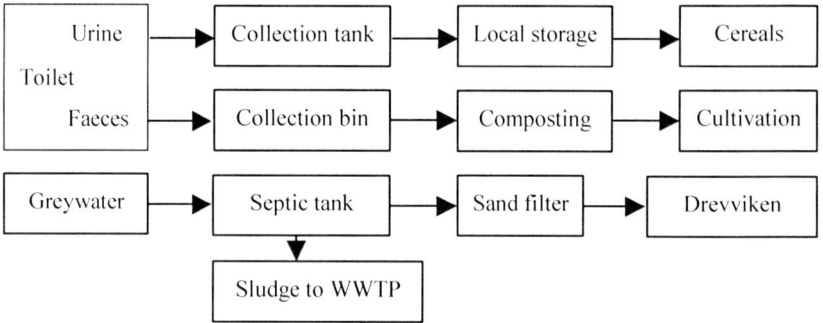

Figure 9-18. Local treatment: Urine reused 11 km away and greywater treated on site. Stormwater discharged to the local receiving water Drevviken.

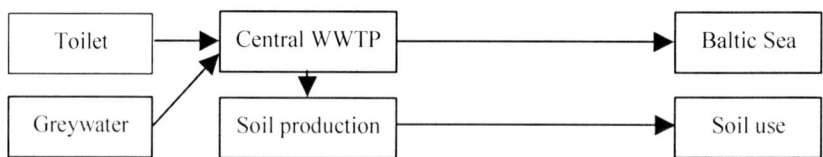

Figure 9-19. Conventional system: Mixed sewage treated at central WWTP. Stormwater discharged to the receiving water.

Studies in the model city Urban Enclave

Measurements of wastewater parameters and hazardous substances in greywater and blackwater in Vibyåsen: The aim was to improve knowledge of the occurrence of hazardous substances and regular wastewater parameters in source-separated greywater and blackwater. The samples were analysed for 105 substances (Palmquist 2004b; Palmquist and Hanæus 2005).

Substance flow analysis (SFA) in Gebers, including hazardous substances and nutrients in greywater, faeces, urine and organic solid waste: Greywater samples were analysed for some 120 parameters and the other fractions for somewhat fewer parameters (Andersson and Jenssen 2002; Jönsson et al. 2005b, Palmquist 2004b; Palmquist and Jönsson 2004).

Environmental systems analysis of an urban enclave in Stockholm: The purpose was to compare source-separating wastewater systems with conventional ones in terms of environmental effects and resource use. This was done using the software package URWARE and LCA methodology (Jönsson et al. 2005b).

Sanitation of faeces using ammonia or chemical oxidizing agents: The aim was to find safe, resource efficient and environmentally friendly sanitation processes for faeces. The sanitising effect of urea and oxidizing chemicals on source-separated faeces were investigated (Vinnerås and Jönsson 2003; Vinnerås et al. 2004).

Faecal contamination and microbial risk assessment of greywater in Vibyåsen: The aim was to improve the knowledge of microbial risk assessments of source-separated greywater. (Ottoson 2003; 2005; Ottoson and Stenström 2003a; 2003b).

Microbial risk assessment of a source-separated system with dry faecal handling: The present Gebers system and a hypothetical local treatment system were compared with a conventional one. Both normal operation and worst-case scenarios were investigated (Jönsson et al. 2005a).

Household aspects of decentralised source-separating wastewater systems: The goal was to describe and analyse the interaction between household routines and a source-separating wastewater system. Theoretical and methodological approaches to the study were developed and implemented (Krantz 2005).

Source-separation of urine and faeces and its technical function: The purpose was to analyse the problems perceived by users of dry toilet systems with and without urine-diversion. This was done using a questionnaire, which was sent to 200 households (Ericsson et al. 2005; Jönsson et al. 2005b).

Chemical risk assessment of wastewater fractions of source-separating wastewater systems: The risks associated hazardous chemicals were evaluated by the multi-step CHIAT procedure, utilising the inherent properties of the chemicals and different possible paths of exposure (Eriksson, E. et al. 2005; Ledin et al. 2005). For details on CHIAT, see Section 3.2.

Box 9-3. The studies in the model city Urban Enclave.

Hygiene and microbial risks

The faecal contamination of source-separated greywater was determined to be 0.040 grams per person and day in the measurements at Vibyåsen (Ottoson 2003). This determination was made using the chemical parameter coprostanol, which gives good estimations of the faecal contamination of source-separated greywater, while the commonly used indicator organism *E. coli* seriously over-estimates this contamination. The growth potential of pathogens, *e.g. Salmonella,* in greywater was found to be small. The microbial risk assessment showed that the reduction of pathogens should focus on viruses and to some extent *Campylobacter;* and also that hygienic safety measures need to be taken when greywater is treated by infiltration, to achieve a 95% safety margin when the local groundwater is used as drinking water (Ottoson 2003; Ottoson and Stenström 2003a).

The hygienic risk assessment of the dry toilet system pointed out that, 1) the emptying of the faecal container, 2) the periodical occurrence of flies in the toilet, and 3) the activities associated with composting of the faecal matter, are the hygienically most risky aspects of the system (Jönsson *et al.* 2005b). These aspects are the main reason for the present and the local treatment systems to be assessed as hygienically more risky than the conventional system.

Another reason is that very few people are exposed to potentially contagious matter in the conventional system. Possible exposure of bathers in the local receiving water Drevviken, to pathogens from the locally treated greywater, led to a somewhat higher hygienic risk for the local treatment system than for the present system. This difference would be far smaller if the local greywater treatment had included an effective chemical precipitation of phosphorus, since this would also reduce the pathogens.

The hygienic risk associated with composting could be eliminated using chemical sanitation instead of composting. An addition of 3% urea-N or 1% peracetic acid kills *Enterococci, E. coli* and viruses (Vinnerås and Jönsson 2003; Vinnerås *et al.* 2004). As treatment with urea also eliminates the risk of regrowing pathogens and can be resource efficient and environmentally friendly, it is an interesting treatment alternative.

Interest in source-separating wastewater systems is often motivated by the advantages of their environmental and resource aspects. The URWARE substance flow software package (Section 2.1) was used to compare the three systems above: the present system, local treatment and conventional system.

Environment

The recovery and reuse of plant-available nitrogen (N), potassium (K) and sulphur (S) proved higher for the source-separating systems than for the conventional one, since these elements are to a large extent lost with the WWTP effluent. Approximately 60 times more N, 16 times more K and 3.5 times more S was recovered and reused by the present system than by the conventional one. The total recovery and reuse of P was essentially the same for the present and the conventional systems, but approximately 25% less for the local treatment one.

Anthropogenic eutrophication is an environmental effect largely caused by wastewater effluents, and its abatement is prioritised. The conventional system was estimated to cause approximately 10 times as much total eutrophication potential (approximately 23 kg O_2 per person and year in the maximum scenario, when both nitrogen and phosphorus contribute) as the present system (2 kg O_2 per person and year, Jönsson *et al.* 2005b). One reason for this was that the model predicted the influent greywater from Gebers to improve the efficiency of the WWTP, through decreased ratios of N:COD and P:COD. This effect has been both predicted and found also by others researchers (*e.g.* Wilsenach and van Loosdrecht 2003, Jernlid and Karlsson 1997). For the local treatment system, the total eutrophication potential caused by water emissions was approximately 2/3 that of the conventional system. However, the total eutrophying emissions were approximately the same, due to larger air emissions of ammonia from the local treatment system.

The energy balances of the source-separating systems were positive, and thus preferable to the negative balance of the conventional system. The present and local systems saved approximately 80 and 100 MJ per person and year, respectively, mainly through savings in chemical fertilisers. The conventional system consumed approximately 30 MJ per person and year: it produced more than twice as much biogas as the present system, but this was more than offset by the use of electricity also being more than twice as much. The use by the systems of the non-renewable resources phosphorus, potassium, sulphur, oil, gas and uranium was weighted according to their abundance and total global use. Compared with the present system, the conventional one used about 8 times, and the local treatment system about 7 times, as much of the weighted non-renewable resources.

Chemical risks

The chemical risks of the systems were analysed using the CHIAT[1] tool. However, input data was lacking for many substances and no heavy metals were included (Eriksson, E. *et al.* 2005; Ledin *et al.* 2005). Consequently, the analysis was supplemented by a literature review covering heavy metals and pharmaceutical substances. Thus, the assessment of the chemical risks associated with the systems is preliminary.

The CHIAT analysis showed that the present and local treatment systems emitted, to arable land, 144 potentially hazardous organic substances coming from urine and faeces. These were mainly pharmaceuticals (50%) and pesticides (25%). In addition, 46 potentially hazardous organic substances were found in the soil produced from greywater sludge. With storm water and treated greywater more than 170 potentially hazardous organic substances was led to the receiving surface water. In the conventional system, approximately 145 potentially hazardous substances were carried in the sewage sludge to the soil produced. Compared with the present and local treatment systems, the number of potentially hazardous organic substances emitted to surface water increased to more than 250, due to the emission of potentially hazardous hydrophilic substances in urine and faeces.

The result of the CHIAT analysis, *i.e.* the number of potentially hazardous organic substances emitted to arable land and receiving surface water, is both thought provoking and disturbing. However, the number of substances may be reduced drastically when limited to priority pollutants among the potentially hazardous substances. For this, the concentrations of the substances need to be known, and further development of the CHIAT tool is also needed.

The supplementary literature review showed that the use of source-separated urine and faeces as fertilisers is consistent with the goal proposed by the Swedish EPA to achieve a mass balance for non-essential substances, including heavy metals, on arable land (Palmquist and Jönsson 2003); however the use of sewage sludge of present quality is not consistent with the goal. The Swedish Medical Drug Agency, in a recent assessment, found it well established that two hormone-active substances, estradiol and etinylestradiol, cause negative effects when emitted to the aquatic environment. Moreover, this was suspected but, due to lack of data, not established for another five substances (SMPA 2004). In the

---

[1]CHIAT = The Chemical Hazard Identification and Assessment Tool. CHIAT consists of five steps: 1) Source characterisation; 2) Identification of recipients, exposure objects and evaluation criteria; 3) Identification of hazards and problems; 4) Hazard evaluation, and 5) Expert assessment. For further details, see Section 3.2.

present and local treatment systems, these substances were emitted to arable soil, which is far better at degrading organic substances than aquatic environments. Thus, the chemical risks established, according to present knowledge, were reduced more by the present and local treatment systems than by the conventional system.

Household aspects and technical function

In discussions on social sustainability, the acceptance of the systems by the users is usually pointed out. However, this does not have the same relevance for Gebers, as the residents themselves chose the system. This study has therefore focused on the interface between the user and the technical installations and on the required household management (Krantz 2005).

Important drivers for most of the residents were an ecological philosophy combined with a global conscience: a conviction that all humans have the same right to the limited resources on earth. Although water is an abundant resource in Stockholm, and water saving as such not a priority, it was considered important not to transport faeces with water suitable for drinking. The dry toilets reduced the use of water and the emission of eutrophying substances. Apart from this, the use of water and potentially polluting household chemicals differed only marginally from that of other residential areas in Stockholm.

The responsibility of each household to empty its own faecal collection bin was strongly emphasised, and few residents would consider emptying their neighbours' collection bin. Faeces were considered intimate and private as long as it could be traced back to the individual or household. In the faecal compost, the faeces lost its identity and thus it was acceptable for this to be managed by a group of residents. The routines for emptying the faecal collection bins have been changed. Due to problems with flies, insufficient drying and the great weight of filled bins, the frequency of emptying has been increased. It now varies from once a month to once a year among the households. When flies appear, the bin is emptied immediately, and this may be repeated several times before the flies disappear. The occasional presence of flies is declared one of the greatest disadvantages of the system; the flies are perceived as unhygienic and disturbing.

Here, the perception of the inhabitants agreed well with the microbial risk assessment of the experts. The residents strive to maintain a certain distance between themselves and the faeces, reducing the exposure to faeces and pathogens and thus the hygienic risk (Krantz 2005). In households including a man, he is generally responsible for emptying the faecal collection bin. Compared with the men, many of the women declared a stronger objection to managing the faeces. Many residents expressed worries about how to empty the collections bins when they grow old, and hoped that this service could be exchanged between neighbours, or bought.

The occasional clogging of the urine pipes was stated as the main problem associated with the source-separated urine. Some users have removed the U-bend (trap) of the toilet altogether, since this is where the clogging first occurs. This may, however, increase both the risk of clogging in other parts of the pipe system and the risk of odours from the urine bowl. Where the pipes have been properly dimensioned and installed with the recommended slope[1], no clogging has been reported, except in the urine trap. Urine pipes should, just as ordinary wastewater pipes, provide good access for inspection and cleansing (Ericsson *et al.* 2005).

Another aspect stated to need improvement was the design of the toilet, to achieve a good source-separation of the urine, especially for women and children. No man considered it a problem to sit down when urinating. Most households considered the toilets with dry faecal collection difficult to clean, largely because no water should enter the faecal bin and cleaning tools adapted to this were lacking. However, the fact that the toilets had fewer odours than ordinary flushed ones improved their hygienic status among the residents. Due to the forced ventilation, there was no odour even when used by several people in succession. However, the fan also makes the system vulnerable; when it fails, the odour soon makes the flat unbearable, if the toilet pipe is not efficiently blocked off.

Poor planning and poor design of the faecal collection bins and the space around them caused most of the serious problems reported by the users of dry faecal collection toilet systems. The bins and their surrounding space should be planned before construction to facilitate ergonomic handling and management. The users were more satisfied with systems that have small collection bins close to the toilet than with large bins in the basement; the probable explanation was that small vessels were emptied more frequently, which made the emptying easier. At the same time, the risks of insects, odours and moisture in the bin, due to carelessness, are reduced, as the bin and the space around it is easy to see and easily cleaned when emptying. Resistant, hygienic and durable materials should be used for all parts of the system. Toilets made of plastics were considered more difficult to clean and were perceived as unhygienic when they got old, partly due to permanent discolouring. The faecal drop pipes should be corrosion-

---

[1]For new installations of urine collection pipes, a pipe diameter of preferably at least 70 mm (absolute minimum 50 mm) is recommended, when the pipes can be easily inspected, cleaned and dismantled. If the pipes cannot be dismantled, *e.g.* underground, a pipe diameter of preferably at least 110 mm (absolute minimum 70 mm) is recommended. The slope should be at least 1%, preferably 2%.

resistant and facilitate cleaning. Information given to the users about the operation, management and maintenance of the whole system, pipes, bins, toilets and reuse of the products is very important and needs to be improved.

9.4.4 Syntheses

Five aspects have been identified as crucial for the long-term operation of the system: responsibility, organisation, comfort, function and information, where function includes reuse of the source-separated and sanitised wastewater fractions.

Responsibility and organisation

The source-separating system at Gebers was designed, constructed, implemented and is now managed by a committed group of residents. The participation of the municipality has been limited to granting permission for the construction and operation of the system. Similarly, several Swedish municipalities have refused to take any responsibility for managing source-separated toilet fractions. This has complicated the introduction of several systems and may risk their operation in the long-term. For source-separating wastewater systems to become more common in the future, the municipalities need to take responsibility for the collection, treatment and re-use of the source-separated fractions. In fact, according to the Swedish Environmental Code, the municipalities already have this responsibility. However, a municipality may allow households to manage their own waste, including the toilet fractions, if this is done in a sanitary and environmentally friendly way. The recycling of nutrients between society and agriculture requires close and well-functioning co-operation between the actors involved: toilet users (the public), municipalities and farmers. Experience shows that the establishment of this trustful co-operation may be the greatest challenge when introducing source-separating toilet systems.

Comfort, function and information

Compared with the conventional system, the systems with urine-diverting double-flush toilets, and where faeces is mixed and co-treated with greywater require only a minor additional effort and inconvenience for the households and building proprietors (Jönsson *et al.* 2000; Johansson, M. *et al.* 2001). Provided that both residents and building proprietors know the urine is put to good use, they are generally well motivated to make the extra effort (*e.g.* Ericsson *et al.* 2005; Kärrman *et al.* 2005b). The function and comfort of well-constructed urine-diverting systems are already sufficient for installation on a large scale, targeting first single-family houses and housing co-operatives. Large-scale implementation is important for improving the system and its components and for decreasing costs.

Wastewater systems with dry faecal collection are not yet well adapted for the regular user. They require considerable improvement in terms of function and comfort before they can be recommended for use in houses with permanent residency of others than highly motivated users. However, as shown by the environmental systems analysis, the environmental and resource aspects of these systems are generally superior to conventional ones.

The social acceptance of source-separating toilet systems is improved if the user has made an informed choice of the system. Informed choice means that the user understands the system better and is often motivated to make an effort to make the system function properly. A visit to a reference facility is a good way to get relevant information for an informed choice. In addition, reference facilities are important for gaining experience of the organisation and management necessary for the proper functioning of the whole system. Good knowledge and information about the system is also a crucial factor for improving system function, reducing its risks and increasing its comfort. A well-written manual is very important. The manual should not only deal with the function of the system, but should also cover its proper management and maintenance, and include a short *do it yourself* section, and a list of spare parts with their suppliers.

City planning and architecture

The introduction of source-separating wastewater systems can substantially increase flexibility in the planning of a city. Since the toilet waste is handled at the source, the demand for a large-scale wastewater system is reduced or eliminated. Drainage water and stormwater still need to be managed, and greywater treated. However, if non-phosphate detergents are used, the treatment can often be limited to the reduction of organic matter (BOD), and pathogens. Thus, sufficient greywater treatment can be achieved by relatively simple technology, especially as the faecal contamination of source-separated greywater is only about one per mille of the faecal content of conventional wastewater (Ottoson 2003). The advantage of placing new residential areas near existent connections to the central wastewater system is reduced. Source-separating wastewater systems with dry faecal management can also increase flexibility in the planning of houses, since the need to place all toilets close to a central wastewater main is reduced. It is up to the architects and the proprietor to use this increased flexibility in the best way.

9.4.5 Conclusions

The most important prerequisite for the dry source-separating toilet system at Gebers was the thorough commitment of the residents. It is unusual to find such a large and committed group of people who are willing to make personal sacrifices, in terms of economy and unpleasant chores, to minimise their contribution to the use of commercial fertilizer, limited resources and to the discharge of eutrophying substances. The interest in the system, shown by scientists and others, has probably increased this commitment.

Source-separating systems with dry faecal management are superior to conventional systems in terms of environmental effects and resource use; so far, however, they are inferior in terms of risk of disease transmission, odour and flies, and thus of social acceptability. Their improvement remains a crucial challenge. This challenge should be possible to meet, as the amount of faeces produced per person and day is only about 160 grams wet weight (53 grams of dry matter) including toilet paper (Jönsson *et al.* 2005a). A well-designed toilet system with dry faecal management would make it possible to build indoor toilets everywhere that improved sanitation is lacking. The need for improved sanitation is immense, as approximately 2.6 million people still lack it.

9.5 INTEGRATED PLANNING IN THE CITY OF UPPSALA

9.5.1 Description of the area and the systems studied

The city of Uppsala, located 70 kilometres north of Stockholm, is Sweden's fourth largest municipality. Although the city is perhaps best known for its 15th century university, it is today also a centre for the pharmaceutical industry. The inhabitants are supplied with drinking water from groundwater, complemented with surface water infiltrated into an esker. Two new drinking water plants were under construction during the years 2004 and 2005. These plants are equipped with a softening process to minimise calcium deposits in installations, and also to reduce the corrosion of copper pipes, and thereby copper concentrations in sewage sludge. The wastewater is transported through a duplicated sewer system (*i.e.* wastewater and stormwater separated) to Kungsängsverket wastewater treatment plant (WWTP). The treated wastewater and a part of the stormwater are discharged to the Fyrisån River. Stormwater from other areas is discharged directly to the recipient. The Fyrisån flows into lake Mälaren, which is the source of water for 1.5 million people. A benefit from the separation of wastewater and stormwater is that the sewage sludge generated in the WWTP is not contaminated with stormwater pollutants, *e.g.* heavy metals from road areas.

A strategy for the reuse of nutrients from sewage sludge was formulated in the year 2000. For societal economic reasons, the main direction in this strategy is a continuation of conventional treatment of wastewater and the use of sewage sludge as a fertiliser in agriculture (Uppsala municipality 2000). In the past, a large proportion of the sludge from Uppsala was used in agriculture, until the food industry banned the fertilising with sludge in the late 1990s, for fear of chemical risks in the food produced. In Uppsala, however, the planners expect that the agricultural use will become feasible again when the sludge quality is improved and public trust restored. Currently, the sludge fulfils all requirements stated by the Swedish Environmental Protection Agency (SEPA), except for too high concentrations of copper.

Uppsala was selected as the concluding project of Urban Water's model city programme, in which tools, models and results from the other model cities were synthesised. The project consisted of two parts. This chapter deals mainly with the strategic analysis and decision support of wastewater management projected to the year 2020, with emphasis on the handling of sewage sludge. The other part, about stormwater management, is summarised at the end of this section.

The systems analysis of wastewater management was limited to the centre of Uppsala, which in 2003 consisted of 135,000 people living in 65,000 flats and, according to the city plan (Uppsala municipality 2002), was predicted to be 165,000 people in 81,000 flats in the year of 2020 (80% living in flats and 20% in detached houses).

The project on strategic analysis and decision support of wastewater management was carried out in cooperation between employees of Uppsala municipality and researchers from Urban Water, from hereon called the project group. The project consisted of the following steps:

1.   Defining the goal and scope of the study,
2.   Formulating alternative solutions,
3.   Comparing the alternatives,
4.   Testing strategic decision-making, supported by a multi-criteria decision aid.

The formulation of the goal and of the alternatives (steps 1 and 2) was carried out in a series of three meetings of the project group over a period of three months. The purpose of the study was to select a strategy for sustainable wastewater management for to year 2020. Four system alternatives were elaborated:

1.   Business as usual, i.e. no recycling of nutrients;
2.   Capture of nutrients at the source by urine separation;

3. Direct agricultural reuse of sewage sludge; and
4. Capture of nutrients at the treatment plant by supercritical water oxidation.

The researchers analysed the chosen alternatives, using the Urban Water tools (step 3). Finally, an authentic testing of the strategic decision-making process (step 4) was carried out in a meeting with the directors of all departments of Uppsala municipality that had a stake in the matter.

Alternative 1: No recycling of nutrients, sludge as "construction soil

In the business as usual alternative, all wastewater is conveyed to the WWTP and treated mechanically, biologically and chemically (*i.e.* by phosphate precipitation). The sludge is composted and stored until needed as "construction soil", for example to cover landfills (Figure 9-20).

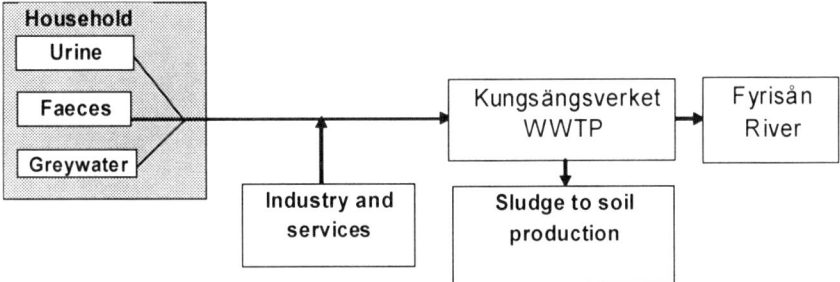

Figure 9-20. Alternative 1: Business as usual, no recycling of nutrients; utilisation of sludge as "construction soil".

Alternative 2: Capture of nutrients at the source by urine separation

In Alternative 2 (Figure 9-21), urine-separating toilets are introduced in some of the flats in Uppsala. Three categories of residential areas are assumed to be equipped with this type of toilet:

- New residential areas to be built during 2008 – 2020;
- Transition areas where former municipal or commercial buildings are to be converted into flats during 2008 – 2020; and
- Residential areas that undergo major renovations during 2008 – 2020.

All in all, 17,500 flats (7,500 newly built or converted, and 10,000 renovated) would have urine separation. From the water-flushed separation toilets, urine and flush water are piped to collective storage tanks that serve 20 households each. The tanks are emptied four times a year, and the urine transported by truck to a seasonal storage site close to the point where it will be spread on the fields.

The remaining wastewater, as well as all the wastewater from flats with conventional toilets, is handled as in Alternative 1.

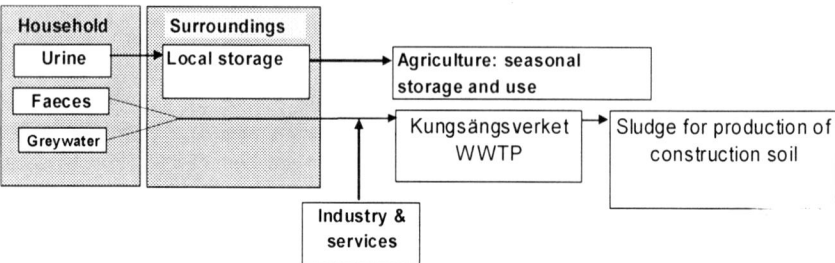

Figure 9-21. Alternative 2: Capture of nutrients at the source by urine separation.

Alternative 3: Direct reuse of sewage sludge in agriculture

Alternative 3 is the same as Alternative 1, business as usual, except for the management of sewage sludge. Here, the sludge is used as fertiliser without further treatment (Figure 9-22). This alternative is not feasible today without a change in policy, or additional pollution control measures.

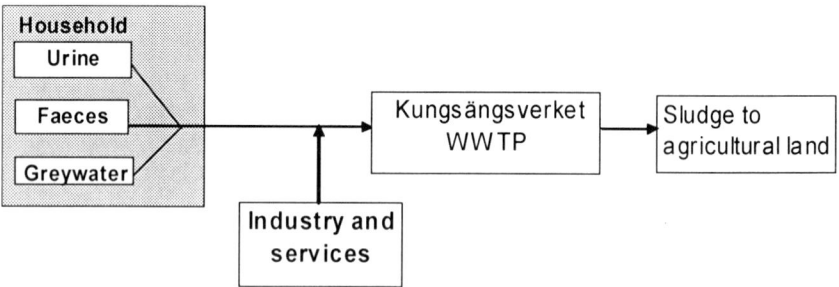

Figure 9-22. Alternative 3: Direct reuse of sewage sludge in agriculture.

Alternative 4: Capture of nutrients at the WWTP by supercritical water oxidation

In the fourth alternative (Figure 9-23), phosphorus is recovered from the sludge by means of the AquaReci process, which builds on supercritical water oxidation. The sewers and WWTP are the same as in Alternative 1. So far, AquaReci has been tested only on pilot-scale. Based on that experience, the footprint of an AquaReci plant in Uppsala would be around 300m^2. The phosphorus-rich product would be transported to a fertiliser factory for further refinement, and the residual ash disposed of.

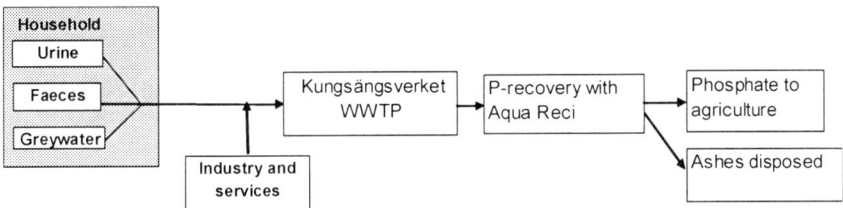

Figure 9-23. Alternative 4: Capture of nutrients at the WWTP by supercritical water oxidation.

9.5.2 Investigations and systems analysis

In a series of meetings, representatives of the city planning department, the environmental department, the water and wastewater department, and from the Urban Water programme, formulated a list of case-specific sustainability criteria. The criteria were: health and hygiene, environment, organisation and implementation, household aspects, economy and technical function. During the meetings, a set of 12 subcriteria was developed (Table 9-5). An internal report was produced by using the stated tools and methods (Kärrman *et al.* 2005c).

Table 9-5. Sustainability criteria for wastewater management in Uppsala.

Main criteria	Subcriteria	Tools for analysis
Health and hygiene	1. Risk of infection	Microbial Risk Assessment
Environment	2. Eutrophication 3. Nutrient recycling 4. Toxic compounds to soil 5. Energy use	Urban Water Research model (URWARE)
Organisation & Implementation	6. Value coalition between crucial actors 7. Sphere of action: legislative political support 8. Presence of "policy entrepreneurs"	Practical guidelines
Households	9. Confidence in the environmental benefits of a system 10. Trust in actors involved	Practical guidelines
Economy	11. Annual costs	Urban Water costing model
Technical function	12. Technical function	Technical risk analysis

The internal report

Health and hygiene

The health and hygiene aspects represented by the subcriterion risk for infection were evaluated by microbial risk assessment (MRA). The results show that the operational staff at the WWTP had the largest risk of infection; hence, the closed sludge treatment in Alternative 4 (supercritical water oxidation), without direct exposure of people, was advantageous. In total, 26 infections per year (12 of which lead to cases of disease) were expected in Alternative 4, compared with 40 to 45 infections per year (18 - 20 cases of disease) predicted for the other three system alternatives. It should be noted, however, that the staff at the WWTP might be immune to many pathogens, thanks to earlier infections.

Environment

The four environmental subcriteria were assessed with URWARE. Phosphorus was selected as the most critical eutrophying substance. The URWARE simulations showed, however, that the phosphorus discharge was similar in all of the alternatives studied, with only a small advantage for urine separation.

Regarding nutrient recycling, no scenario was able to meet the long-term goal of the Swedish EPA (SEPA), that all macronutrients (N, P, K, S) should be recycled. The SEPA short-term goal requires that 60% of phosphorus contained in wastewater be recycled by the year 2015, half of which to agriculture, and half to other types of cultivation (*e.g.* parks). Two of the alternatives, 3) direct reuse and 4) supercritical water oxidation, would comply with the short-term goal.

The fourth subcriterion, toxic compounds to soil, was represented by the flow of cadmium to arable land. By using URWARE, it was shown that all system alternatives, except for Alternative 3 (direct reuse), resulted in very small amounts of cadmium being transported to arable land. When reused, a major share of the cadmium that entered the wastewater was also transported to arable land. However, also this system alternative complied with the regulatory limit for cadmium in sewage sludge.

All of the four alternatives involved both energy use and energy recovery. Electricity is used for pumping and treatment processes, and fossil fuels for transports. Energy is recovered through generation of biogas. The direct reuse (Alt. 3) consumed the least energy, while the alternative with supercritical water oxidation was the most energy demanding.

Organisation and implementation
Municipal employees and politicians who work with the urban water system in Uppsala were interviewed about their opinions on decision-making and sustainable development. In total, 19 people were interviewed in 17 interviews. Although the interviewees were not asked specifically about the technical solutions of the four system alternatives investigated, it was concluded that a value coalition between crucial actors existed for Alternatives 1 and 3, the conventional solutions. Only one politician was positive to urine separation (Alt. 2); all of the employees of municipal utilities were sceptical. A wait-and-see policy was adopted towards the supercritical water oxidation process (Alt. 4), which means that the technology may be implemented when it has proven successful in full-scale applications elsewhere. The conventional system alternatives were regarded as having action sphere, *i.e.* legislative political support.

An uncertainty with Alternative 1 was, however, the possible introduction of legislation to require phosphorus recycling. In Alt. 2, the introduction of urine-separating toilets may be difficult, legally, to require of home owners. Furthermore, the legislation about spreading urine in agriculture is unclear. In the event that phosphorus recycling becomes legally binding, supercritical water oxidation is a probable add-on process to the existing technical infrastructure. The greatest problem in Uppsala seemed to be the localisation of a facility for this. Regarding the subcriterion, the presence of policy entrepreneurs, it was found that Alternatives 2, 3 and 4 lacked such initiators and implementers, while such people were already active in all parts of Scenario 1.

Households

Information about household aspects was compiled from literature reviews and from three meetings with 10 people in the Gottsunda residential area, which was one of the candidates for the introduction of urine separation (Kärrman *et al.* 2005c). The subcritrerion of Confidence in the environmental benefits is relevant only for Alternatives 2 and 3; it would require a change in routines. A successful implementation of urine separation demands an active role of the households, mostly by using the toilet in a correct way. If sewage sludge is to be spread on arable land without additional treatment, the diffuse pollution by heavy metals and hazardous chemicals needs to be reduced. How the residents could contribute to this, and to what extent it would actually improve the sludge, was unclear to them. Alternatives 1 and 4, on the other hand, do not require any interaction with the households.

The trust in the municipal actors was higher among the residents than among the property owners. Both groups believed that systems where everyone, including their neighbours, must fulfil their part of the collective responsibility are difficult to implement successfully. There were also doubts among the residents about whether human excreta should be spread on arable land at all, for hygienic reasons.

Economy

The Urban Water costing model (Section 4.1) was applied to estimate the annual cost for the new construction of the technical infrastructure in the four system alternatives (Kärrman et al 2005c). Since all investment costs were calculated for building entirely new systems, the differences between them were relatively small. The direct reuse alternative (3) had the lowest annual cost. Urine separation was the most expensive, though only 9% higher than the cheapest one.

Technical function

The term technical function included both the households' requirement of reliable technology, and the risk of technical malfunction for the organisation that owns and operates the municipal wastewater system. For the households, reliable technology had the highest priority, which included aspects such as toilets that do not leak and drainpipes that do not become blocked. It was obvious that people who have experienced these kinds of problems develop a negative attitude towards water and wastewater utilities, even if the responsibility to avoid or repair the problem was actually their landlord's.

It was concluded that Alternative 1, business as usual, is a technically robust system for households as well as for the municipality. Urine separation involves potential obstacles not only for the household, with cleaning the toilet and blocked pipes, but also for house owners (landlords), who are responsible for emptying the storage tank. Direct reuse (Alt. 3) and oxidation treatment (4) imply somewhat higher technical risks than the business as usual. Direct reuse requires that households not to flush down contaminants, while supercritical water oxidation introduces the workplace safety issue of high pressures and temperatures.

A multi-criteria decision support process

A report prepared by the project group was presented to the directors of the municipal departments involved, *i.e.* the city planning department, the department of water and sewage works, the department of environment and health, and the department of municipal buildings. In this session, a process for strategic decision-making was tested, using a multi-criteria decision aid, STRAD.

9.5.3 Results: Comparisons of the systems studied

During the decision-making session, the participants placed the alternatives on a graphical scale criterion by criterion. One example is given in Figure 9-24 where the cost subcriterion was assessed for the four alternatives on a relative non-graded scale from high costs to low costs. It should be noted that the group did the placing of the alternatives on the scale in consensus. They had the calculated costs from the internal report available, but could also add own arguments in their assessment.

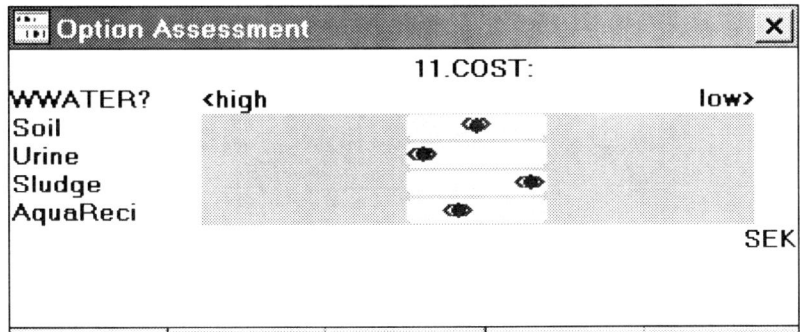

Figure 9-24. Assessing the cost criterion in STRAD by placing the alternatives on a non-graded scale from "low" to "high".

After working through all of the criteria, an overview can be visualised (Figure 9-25). The fields on the right of each criterion represent the magnitude of the span from between the alternative with the lowest cost and the alternative with the highest cost.

Then an overall assessment was made (Figure 9-26). The outcome was that Alternative 1, business as usual, was the most preferable system, followed by Alternatives 3 and 4. Alternative 2) Urine separation received the lowest ranking.

In STRAD, there is no direct weighting between the criteria, *i.e.* all criteria have the same weight initially. The width of the fields for each criteria can however be changed. Enlarging the width of the field means more "weight" on that specific criterion and diminishing the width means less "weight".

A sensitivity analysis was carried out by changing the width of the criteria in the overall assessment. When the subcriterion Risk of infection was assigned a lower priority, and the cost parameter a higher one, the direct reuse of sludge ranked the highest (Figure 9-27).

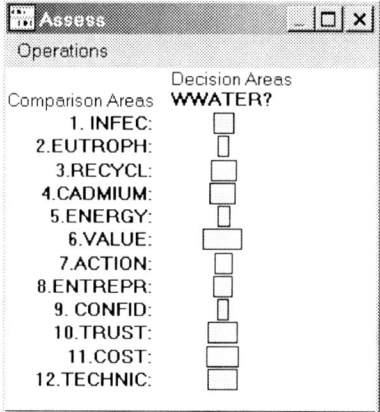

Figure 9-25. An overview of the subcriteria, in STRAD called "Comparison areas".

	< Less preferred	More preferred >
Soil		⬡
Sludge		⬡
Aqua Reci	⬡	
Urine sep	⬡	

Figure 9-26. Overall assessment with STRAD.

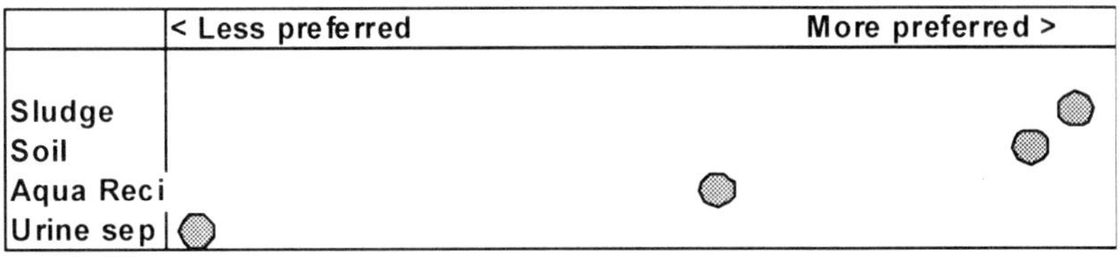

	< Less preferred	More preferred >
Sludge Soil Aqua Reci Urine sep		

Figure 9-27. Ranking of system alternatives after adjusting the weighting of two subcriteria, *i.e.* less priority to "risk for infection", and more to "cost".

In a second sensitivity analysis (Figure 9-28), only the environmental subcriteria were taken into account. Here, Urine separation ranked the highest.

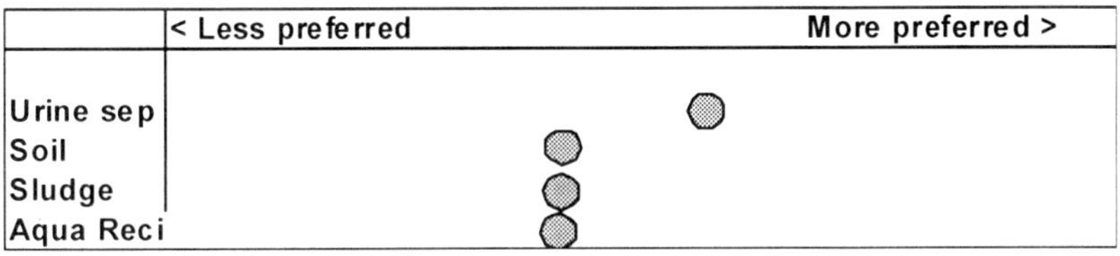

	< Less preferred	More preferred >
Urine sep Soil Sludge Aqua Reci		

Figure 9-28. A ranking taking into account only the environmental subcriteria.

9.5.4 Lessons learnt

The participants in the STRAD session thought it a positive experience; they found the approach useful for municipal planning, since it structured the decision-making process and improved the dialogue and exchange of knowledge between the actors. A proposal from one participant was to use STRAD early in the process for screening, *i.e.* the selection of a few alternatives for deeper analysis, from a rather unmanageable smorgasbord of initial suggestions. In real decision-making processes, a lack of time often forces the planners to reduce the number of alternatives considered, without having a rational procedure for setting priorities.

A key strength of the approach was the structured procedure, by which arguments could much more easily be traced back than in simply an open discussion. It was considered an advantage to deal with the criteria one by one. The imbalance in the number of subcriteria under each main criterion complicated the assessment, which caused some participants to question the output of the model.

9.5.5 Systems analysis of stormwater management

The systems analysis of stormwater management in Uppsala aimed to:

> 1) quantify the discharges of a selection of pollutants from stormwater and to assess variations;
> 2) identify the sources of these pollutants in general terms; and
> 3) analyse, in a couple of catchment areas, potential alternatives for improved stormwater management and beneficial use of stormwater (Ahlman *et al.* 2005b).

A substance flow analysis was carried out using the sewer system model SEWSYS. The SEWSYS simulations showed that there is a potential for constructing stormwater ponds in the catchment area, to prevent considerable amounts of copper, zinc and lead from entering the river. Another potential measure to reduce discharges of heavy metals would be to decrease the area of roof surfaces made of copper, as well as galvanised surfaces of building materials in general.

10

Literature

10.1   URBAN WATER REPORTS

Apart from the publications in international journals and at conferences, the results of the programme have been documented in a publication series by Urban Water, Chalmers University of Technology, Göteborg, Sweden, under ISSN 1650-3791. Most of the reports are available for download in pdf format from www.urbanwater.org. Apart from the reports in English language listed below, additional reports are available in Swedish from the same site. Furthermore, some reports were published in the publication series of the Swedish Water and Wastewater Association (Svenskt Vatten). These are available as "VA-Forsk rapport" from www.svensktvatten.se.

Urban Water reports in English language.

Available from www.urbanwater.org

No.	Authors	Title
2001:3	Strandberg, L., Johansson, M., Palmquist, H. and Tysklind, M.	The flow of chemicals within the sewage systems – possibilities and limitations for risk assessment of chemicals
2004:1	Urban Water	Chemical and microbial risks associated with wastewater systems *(workshop proceedings)*
2004:3	Hellström, D., Hjerpe, M., and van Moeffaert, D.	Indicators to assess ecological sustainability in the Urban Water sector - An assessment of the Urban Water programmes' criteria and indicators for analysis of ecological sustainability of urban water supply and wastewater systems
2004:4	The Urban Water Research school	Sustainable Urban Water Management in International Mega-Cities. Experiences from Study Trips to Cairo, Kolkata and Tokyo.
2004:6	Czemiel Berndtsson, J.	Beneficial use of stormwater: a review of possibilities
2005:5	Jeppsson, U., Baky, A., Hellström, D., Jönsson, H. and Kärrman, E.	The URWARE wastewater treatment plant models
2005:6	Jönsson, H., Baky, A. Jeppsson, D., Hellström and Kärrman, E.	Composition of urine, faeces, greywater and bio-waste - for utilisation in the URWARE model
2005:7	Ashbolt, N.J., Petterson, S.R., Stenström, T.A., Schönning, C., Westrell, T. and Ottoson, J.	Microbial Risk Assessment (MRA) Tool
2005:11	Ahlman, S. and Svensson, G.	SEWSYS - a tool for simulation of substance flows in urban sewer systems

10.2   Ph.D. THESES

In the following, a short summary of the Urban Water Ph.D. projects is given. The theses are available from www.urbanwater.org, though, for reasons of copyright, without any of the appended papers. A printed copy that includes the appended papers can be ordered from the respective institutes.

Biofilms and microbial barriers in drinking water treatment and distribution

Jonas Långmark, Department of Land and Water Resources Engineering, Royal Institute of Technology (KTH), Stockholm. Defended 8 December 2004.

The primary objective of conventional drinking water treatment and distribution is to deliver to the consumer water that is both aesthetically pleasing and does not constitute a human health risk. To achieve this, water utilities employ a range of physical (*i.e.* sand and membrane filtration) and chemical (*i.e.* flocculation and disinfection) barriers in order to reduce the numbers of microorganisms as well as the nutrients that may support their growth within biofilms. In this thesis, biofilms and microbial barriers in water treatment and distribution were therefore examined. The development of biofilms within artificial recharge was investigated in pilot column at Norsborg waterworks in Stockholm. The proportion of active bacteria, measured as numbers of EUB338-positive cells relative to the total number of bacteria enumerated by total direct counts, decreased with time. Through the addition of nutrients however, two to three times more bacteria were able to be active (measured by increase in activity after activation with additional nutrients). By extracting the recalcitrant hydrophilic and hydrophobic fractions of humic substances it was possible to assess the microbiological response to those compounds. It was shown that bacteria more firmly attached to the sand grains preferred the hydrophobic fraction whilst more loosely-associated bacteria preferred the hydrophilic one. The amount of easily degradable matter in raw water (measured as assimilable organic carbon) was generally low. Biofilms were investigated by two different methods for extraction and analysis of microorganisms. Glass slides introduced into the sand material were dominated by α-Proteobacteria, and underestimated loosely-associated bacteria whilst extracts from sand were dominated by γ-Proteobacteria, and also caused variations due to the extraction method employed The barrier function of biofilms was investigated in biofilters, also fed with raw water from Gothenburg. The focus here was on particle removal in size-intervals of 1-15 μm (protozoa) and 0.4 – 1 μm (bacteria). In both size fractions, autofluorescent microalgae, which were naturally occurring in raw water, were also enumerated in parallel. Their removal was 60-90%. In parallel, defined amounts of fluorescent hydrophilic and hydrophobic microspheres (1 μm) were added. They showed a reduction of hydrophobic spheres by 98% and hydrophilic ones by 86%. Removal of viruses was determined by adding a defined dose of bacteriophages and gave lower reduction values of 40 – 61%. Both naturally-occurring particles in defined size intervals and added particles or organisms were shown to provide a clearer picture of barrier function than usually performed measurements of turbidity. Efficiency of chemical treatment against viruses was also measured in a pilot-plant in Gothenburg. It was shown that commonly-used MS-2 bacteriophages were much more sensitive than φX174 bacteriophages. Reduction of MS-2 over the entire chemical step (when added after dosing of chemicals) was 5-log10 whilst φX174 was reduced by 1-log10. The latter was shown to be a more conservative model for virus removal. The effects of different steps in the chemical precipitation showed that the primary dosage of chemicals and the development of flocs had great importance for the assessment of removal efficacy. When added before the dosing of chemicals, reduction of φX174 and MS-2 was $3.8\text{-}\log_{10}$ and $6.2\text{-}\log_{10}$, respectively.

The establishment of biofilm within a distribution system was followed in a 1000 metre long pilot-plant (with parallel lines) at Lovö waterworks as well as in two of Stockholm's main distribution systems (Nockeby and Hässelby). The pilot-plant was shown to satisfactorily represent processes within the distribution systems. The development of biofilms was slow, producing thin biofilms over a one to two month periods. Numbers of bacteria were generally in the range of $10^4 - 10^5$ per cm^2, which is lower than shown in other earlier investigations. The implementation of primary ultra violet (UV)-treatment in place of chlorination (both being chloraminated prior to distribution) did not considerably change the numbers of bacteria in biofilms. No significant difference could be seen between the system that had UV-treatment as a primary treatment step, and the system that was chlorinated over the whole period. Chlorine residuals were generally low at the distal parts of the distribution systems. Naturally-occurring protozoa were present in distribution systems in numbers ranging from 280 – 3500 protozoa per cm^2. Protozoa may play a significant role as predators of biofilm bacteria, however they can also act as protection for bacteria against external influences *i.e.* disinfection. Should sudden contamination of a distribution system occur, biofilm can provide protection and act as a site for potential regrowth of introduced microorganisms. Biofilms developed in the pilot-scale that represented water from different distances from waterworks were exposed to fluorescent microspheres, (hydrophobic and hydrophilic, 1 μm) legionellae (as a model for opportunistic bacterial pathogens) and bacteriophages (human enteric virus model) in order to determine their accumulation and persistence within the biofilm, and release to the bulk water. It was shown that introduced model organisms were released continuously, primarily through desorption, and additionally through the influence of disinfection and activity of protozoa. Desorption was also assessed in a laboratory experiment under laminar and turbulent flow. Laminar flow conditions that were representative of a distribution system gave a slow and continual release of individual cells, whilst turbulent conditions detached larger aggregates. In conclusion, based on this work an increased understanding was gained both of barrier functions at the different steps of water treatment, their effects on overall biofilm dynamics and structure and the role that biofilm plays within the drinking water system itself.

Hazardous Substances in Wastewater Management

Helena Palmquist, Dept of Civil and Environmental Engineering, Division of Sanitary Engineering, Luleå University of Technology, Luleå. Defended 10 December 2004.

The extensive use of materials and substances in society causes diffuse source emissions that lead to uncontrolled spreading of hazardous substances, largely channelled via wastewater systems, to the surrounding nature. Complex mixtures of substances appear in wastewater as a result of use, wear, and corrosion of goods (*e.g.* pipes, taps, carpets, furniture) as well as the use of household chemicals from doing the laundry and dishwashing and the use of pharmaceuticals and personal care products. As many as 30,000 substances regarded as everyday chemicals are regularly used in households, implying that the flows of hazardous substances in wastewater systems are not only a complex issue for wastewater management, but for society as a whole. As a part in analysing the flows of hazardous substances in wastewater systems, domestic wastewater fractions (greywater, urine, and faeces) were chemically characterised through full-scale field samplings at the source separating domestic wastewater systems Vibyåsen and Gebers with respect to a selected number of hazardous substances. Data on the characteristics of wastewater fractions were essential to improve the prerequisites for performing substance flow analysis and chemical risk assessment of the wastewater systems. The mass flows of hazardous metals from households emerged in similar quantities in the greywater and toilet fractions. However, ratios of hazardous metals to phosphorus and nitrogen were significantly lower in the urine than in the faecal matter and greywater. The mass flows of organic hazardous substances from households were mainly searched for in the greywater, resulting in 50-60% of the 81 measured substances being found, with representatives from all of the substance groups investigated. Of the 72 measured organic hazardous substances, 36% were found in the blackwater at Vibyåsen. However, it was not possible to exactly identify their specific sources as the mass flows of organic hazardous substances derive from diffuse household sources like everyday activities (laundry, cleaning, etc.), the wear of things such as pipe material and interior fittings, and from airborne deposition. The input of organic hazardous substances to urine and faeces occurs mainly via the excretion of, for instance, pharmaceuticals, pesticides, and food additives. Other examples of relevant pathways are when emptying a scouring pail and throwing in cigarette butts, snuff, etc., into the blackwater via the water closet. Based on a number of recent measurements (including Vibyåsen and Gebers) a proposal for new Swedish design values for nutrients (*e.g.* phosphorus and nitrogen) and seven heavy metals (*e.g.* copper and cadmium) in household wastewater fractions was put forward. However, the consumption patterns of society changes over time, and with it, so do the wastewater characteristics. Therefore, design values should be used with good judgement and require regular updating, assumingly each 5th to 10th year. A possible management approach was suggested to interpret and compare different wastewater systems, and to serve to find out if and how much the flow of hazardous substances can be stopped, diverged, or transformed at the source or during transport throughout the wastewater system. The barriers approach was proposed as a tool on a conceptual level (a way of thinking) as an attempt to support a shift in perspectives by combining a traditional end-of-pipe perspective with more systems-oriented perspectives, thereby linking the use of resources and the spreading of hazardous substances to their underlying causes and driving forces (*i.e.* consumption and lifestyle) rather than only focusing on the emissions. Organisational and behavioural barriers, system design, process barriers, and optional recipients were suggested, implying that various kinds of measures are needed in the management of hazardous substances to achieve a change in direction towards sustainability.

Microbial risk assessment and its implications for risk management in urban water systems

Therese Westrell, Department of Water and Environmental Studies, Linköping Studies in Arts and Science, Thesis no. 304, Linköping University, Linköping. Defended 10 December 2004.

Infectious diseases can be transmitted via various environmental pathways, many of which are incorporated into our water and wastewater systems. Quantitative microbial risk assessment (QMRA) can be a valuable tool in identifying hazard exposure pathways and estimating their associated health impacts. QMRA can be applied to establish standards and guidelines and has been adopted by the World Health Organisation for the management of risks from water-related infectious diseases. This thesis aims at presenting a holistic approach for the assessment of microbial health risks in urban water and wastewater systems. The procedure of QMRA is presented, together with the data collected for the case studies, and the results are discussed in a risk management framework.

Decentralised drinking water treatment with membranes was shown to be competitive with centralised conventional treatment regarding environmental impacts and health. To attain sufficient die-off of pathogens in order to reduce risks to acceptable levels, facilities that permit the long-term storage of locally collected faeces are required. Issues of operation and management are likely to determine the health risks in decentralised systems. While failures in distribution are more likely to result in detectable waterborne disease outbreaks, the number of people at risk of becoming infected with pathogens passing normal treatment, calculated on a yearly basis, can be larger. Site-specific pathogen monitoring of source waters was identified as an important factor for the accurate estimation of risk. Noroviruses, an emerging waterborne pathogen, were shown to have fluctuating concentrations in surface water, with significant peaks during the wintertime. Time series analysis has potential as an early warning system if complemented by regular monitoring to discriminate peaks from random fluctuations. Groups already sensitive to infection, *i.e.* the elderly, the sick and children, were shown to consume higher volumes of cold tap water than the rest of the population, which may call for special attention in the risk management of drinking water systems. Microbial health risks associated with the handling and reuse of wastewater and sludge were shown to be successfully addressed within the management system Hazard Analysis and Critical Control Points (HACCP). Most exposure points identified could be controlled through easy measures.

Biological Pre-filtration and Surface Water Treatment – Microbial barrier function and removal of natural inorganic and organic compounds

Gerald Heinicke, Water Environment Transport, Chalmers University of Technology, Göteborg. ISBN: 91-7291-561-7. Defended 1 February 2005.

Waterworks in Sweden that apply conventional chemical surface water treatment are facing a number of challenges, including changes in raw water quality and demands for improved particle removal. The chief objective of this work was to evaluate biological pre-filtration for typical Swedish surface water, regarding the removal of natural organic matter, particles, iron and manganese, and taste and odour compounds. Pilot-scale experimental work included the investigation of biofilters fed directly with surface water, using non-adsorptive expanded clay and partly exhausted granular activated carbon. Process combinations with conventional chemical treatment and nanofiltration were investigated.

Biological pre-filtration decreased the load of particles and biodegradable organic matter to subsequent treatment processes. Peak loads of added particles were equalised by high initial retention followed by a slow release of attached particles. In chemical treatment with pre-filtration, the removal of μm-size particles became less dependent on the post-sedimentation rapid filters. The experimental study contributed with data on the microbial barrier function of chemical surface water treatment under Swedish conditions with regard to particle-, bacteria- and virus removal.

Simple rapid media filtration pre-treatment of surface water caused fast pressure drop development in nanofilter membranes. Biofiltration moderated the increase in pressure drop in comparison to rapid filtration. Destructive analysis of the nanofiltration elements was performed to study the fouling layer on the membrane.

Beneficial use of storm water – Opportunities for urban renewal and water conservation

Edgar Villareal, Department of Water Resources Engineering, Lund University. ISBN 91-628-6277-4. Defended 4 March 2005.

This thesis investigates the beneficial use of stormwater in Swedish urban environments. The focus has been on open stormwater systems for source control, as well as rainwater collection systems. To analyse the implementation of open stormwater systems, two case studies are presented: Augustenborg in Malmö and Bäckaslöv wetland in Växjö. The retention and detention effects of green-roofs have been also studied by means of experimental work. Ringdansen, a residential area in Norrköping, is used to discuss rainwater collection for urban applications. The aim of the Augustenborg study was to determine the effect of disconnecting impervious areas from a combined sewer system in favour of a new open stormwater system. The open system, which consists of a combination of structural best management practices (BMPs), was assessed by comparing synthetic hydrographs for the ½, 2, 5 and 10-year design storms assuming wet and dry initial conditions. It was found that the BMPs are able to lower the total runoff from the area and successfully attenuate storm peak flows for all the recurrence intervals considered. Bäckaslöv wetland in Växjö, is shown as an example of a multi-purpose urban stormwater facility. The wetland system has been the focus of studies into water quality and hydraulic processes, as well as surveys of plant and bird populations. Apart from stormwater management, the system benefits include urban renewal, habitat creation and the provision of an outdoor laboratory. In order to estimate the response of green-roofs to individual rain events, precipitation and runoff data from controlled experiments (with dry and wet initial conditions) on a sedum album green-roof were analysed by means of linear programming. The results indicated that slope does not influenced retention volumes. Under dry initial conditions, stormwater can be retained and detained, whereas under wet initial conditions only detention is possible. To further investigate the detention effect, a comparison between the response of an impervious roof and a green roof was carried out. Green-roofs can be implemented in combination with other BMPs to successfully detain stormwater. The Ringdansen study showed that rainwater collection systems are efficient elements in Swedish urban environments. They contribute to water saving by supplying water for domestic use where potable water is not essential. Rainwater collection systems are also shown to be important elements of stormwater management systems.

Matter that matters. A study of household routines in a process of changing water and sanitation arrangements

Helena Krantz, Department of Water and Environmental Studies, Linköping Studies in Arts and Science, Thesis no. 316, Linköping University, Linköping. ISBN: 91-85297-65-8. Defended 1 April 2005.

Our society has changed, but the urban water and sanitation system of today is roughly the same as it was 100 years ago. The system is designed for, developed from and sustained by human activities, and has since its introduction affected household patterns of routine activities. The urban water and sanitation system is now being criticised for not being sustainable due to excessive material, energy and chemical use, and failure to recycle and reuse resources. Altering household practices is perceived as one important step towards improved sustainability.

In this study, two changes in water and sanitation arrangements at the household level are analysed: individual meters for volumetric billing of hot and cold water, and dry toilets with separate collection of urine and faeces. These arrangements increase system transparency, and their proponents believe that the arrangements enhance resource recycling and/or resource savings. However, success in this regard can only be achieved if accompanied by appropriate household routines. The extent to which such appropriate routines come about and why (not) is the focus of attention in this study; the aim is to describe and analyse the interaction between householder routines and changes in water and sanitation arrangements.

This study takes as its starting point household everyday life. A methodological combination of time-diaries, interviews, physical measurements and simple observations is developed and implemented in two cases; the housing area Ringdansen with flats (volumetric billing) and the collective Gebers based on an ecological way of life (dry toilets). The theoretical approach is developed from time-geography and culture analysis. The methodological and theoretical approaches have proven useful and can be developed further.

Household responded differently to the volumetric billing in Ringdansen, but in general, no sweeping routine changes took place in the households. A comparison of average total water usage per household (at an aggregated level) between the two cases, showed no significant difference. Water-use routines are also similar in the two areas, even though variations appear between households. There seems to be a socio-culturally defined lower limit for water use, regarded as necessary for maintaining sufficient standards of cleanliness and comfort, irrespective of the influence of ecological or economic incentives. Differences in household composition, built-in technical arrangements and existence of a garden (Gebers) explain the differences in hot and cold-water usage between the two areas. The dry toilet was shown to have a decisive impact on toilet disposal routines; only biodegradable waste products are thrown into it and the cleaning agents are environmentally friendly. The change in technical arrangement, in combination with household responsibility for maintenance and knowledge of that faeces and urine being returned to nature, resulted in environmentally friendly toilet disposal routines that reach beyond the 'good' routines evolving from environmental concern. The relationship between changes in water and sanitation arrangements and householder routines may be expressed as follows: an extensive change in arrangements, either technical/physical, organisational and/or economical, results in more radical routine changes, and more so if combined. However, the improvement as regard ecological sustainability is conditional on what is socio-culturally accepted – social sustainability.

Treatment of domestic wastewater using microbiological processes and hydroponics in Sweden

Anna Norström, Department of Biotechnology, Royal Institute of Technology (KTH), Stockholm. ISBN: 91-7187-030-0. Defended 18 May 2005.

Conventional end-of-pipe solutions for wastewater treatment have been criticized from a sustainable view-point, in particular regarding recycling of nutrients. The integration of hydroponic cultivation into a wastewater treatment system has been proposed as an ecological alternative, where nutrients can be removed from the wastewater through plant uptake; however, cultivation of plants in a temperate climate, such as Sweden, implies that additional energy is needed during the colder and darker period. Thus, treatment capacity, additional energy usage and potential value of products are important aspects considering the applicability of hydroponic wastewater treatment in Sweden.

To enable the investigation of hydroponic wastewater treatment, a pilot plant was constructed in a greenhouse located at Överjärva gård, Solna, Sweden. The pilot plant consisted of several steps, including conventional biological processes, hydroponics, algal treatment and sand filters. The system treated around 0.56-0.85 m^3 domestic wastewater from the Överjärva gård area per day. The experimental protocol, performed in an average of twice per week over a period of three years, included analysis and measurements of water quality and physical parameters. In addition, two studies were performed when daily samples were analysed during a period of two-three weeks. Furthermore, the removal of pathogens in the system, and the microbial composition in the first hydroponic tank were investigated.

Inflow concentrations were in an average of around 475 mg COD/L, 100 mg Tot-N/L and 12 mg Tot-P/L. The results show that 85-90% of COD was removed in the system. Complete nitrification was achieved in the hydroponic tanks. Denitrification, by means of pre-denitrification, occurred in the first anoxic tank. With a recycle ratio of 2.26, the achieved nitrogen removal in the system was around 72%. Approximately 4% of the removed amount of nitrogen was credited to plant uptake during the active growth period. Phosphorus was removed by adsorption in the anoxic tank and sand filters, natural chemical precipitation in the algal step induced by the high pH, and assimilation in plants, bacteria and algae. The main removal occurred in the algal step. In total, 47% of the amount of phosphorus was removed. Significant recycling of nitrogen and phosphorus through harvested biomass has not been shown. The indicators analysed for pathogen removal showed an achieved effluent quality comparable to, or better than, for conventional secondary treatment. The microbial composition was comparable to other nitrifying biological systems. The most abundant phyla were Betaproteobacteria and Planctomycetes.

In Sweden, a hydroponic system is restricted to greenhouse applications, and the necessary amount of additional energy is related to geographic location. In conclusion, hydroponic systems are not recommended too far north, unless products are identified that will justify the increased energy usage. The potential for hydroponic treatment systems in Sweden lies in small decentralized systems where the greenness of the system and the possible products are considered as advantages for the users.

Sustainable Development and Urban Water management – Linking theory and practice of economic criteria

Mattias Hjerpe, Department of Water and Environmental Studies, Linköping Studies in Arts and Science, Thesis no. 322, Linköping University, Linköping. ISBN: 91-85297-87-9. Defended 27 May 2005.

The interest in using criteria and indicators for assessing activities in relation to sustainable development is increasing. This dissertation analyses the potential for using economic criteria for assessment of urban water management in relation to sustainable development. The analysis consists of three parts.

First, to analyse the basis for economic criteria, there is a need to categorise general frameworks, disciplinary theories and practical assessments in order to explore what the economic dimension of sustainable could imply, depending on general assumptions about challenges, goals and means. Consequently, a number of general frameworks, economic theories and urban water assessments were categorised.

Second, based on this analysis, a set of economic criteria was chosen, consisting of maintenance of water infrastructure, affordability, cost-recovery, effectiveness and development potential. For each criterion, one or more indicators are suggested.

Third, these indicators were tested in three cases from Swedish municipalities: introduction of volumetric billing in a low-income apartment area, increased water supply in a growing city and introduction of kitchen waste disposers in a city with a stagnant population.

On the basis of the application, introduction of volumetric billing in a low-income area resulted in deteriorating affordability and effectiveness, whereas cost-recovery improved. Introduction of kitchen waste disposers in a stagnant area was questionable from an effectiveness viewpoint whereas the water infrastructure was well maintained. In the growing city, increased income and population determined the outcome of the affordability, cost-recovery and development potential criteria, which all improved.

The study also found that using economic criteria and indicators for assessment of urban water management in relation to sustainable development requires a continuous balance between the universal and the context-specific, that is, between the criteria and indicators used and the water infrastructure change being assessed. This emphasises that criteria used should relate to all dimensions of sustainable development as well as of the decisiveness of involving actors and other stakeholders in sustainable development assessments.

Comparative analysis of pathogen occurrence in wastewater

Jakob Ottoson, Department of Land and Water Resources Engineering, Royal Institute of Technology (KTH), Stockholm. Defended 2 June 2005.

This project was initiated to fill knowledge gaps on the occurrence of pathogens in different streams of wastewater, *e.g.* greywater and domestic wastewater. The aims were also to measure the removal of pathogens in different treatment processes, conventional and innovative, and correlate the removal to that of common microbial process indicators, such as faecal coliforms, enterococci, *Cl. perfringens* spores and bacteriophages. One study also assessed the correlation between the removal of microorganisms and some commonly measured physico-chemical process indicators. The results can be applied in microbial risk assessments (MRAs) of urban wastewater systems.

Indicators and parasitic (oo)cysts were enumerated with standard methods and viruses with rtPCR. High levels of *Giardia* cysts and enteroviruses were found in untreated wastewater ($10^{3.2}$ and $10^{4.2}$ L^{-1} respectively) indicating high incidences in the society. Noroviruses were also often found in high numbers ($10^{3.3}$ L^{-1}) during winter, but less frequent and in lower numbers ($10^{2.3}$ L^{-1}) during the rest of the year. This temporal variation correlated to the clinical laboratory reporting of noroviruses. A temporal variation was also shown for *Giardia* with significantly lower cyst counts in untreated wastewater during spring. *Cryptosporidium* oocysts were not as numerous in untreated wastewater (5 L^{-1}) reflecting a lower incidence in the society than for the other pathogens during the time of the study. Since temporal variation had a larger impact than spatial, site-specific measurements may not be necessary to perform screening level MRAs of wastewater discharge and reuse. Good data can be found in the literature and corrected for by recovery of the detection method, flow and incidence in the society.

Removal of microorganisms in wastewater treatments varied from 0 to >5.8 log due to process combination and organism in question. Treatment in integrated hydroponics removed microorganism more efficiently than did secondary conventional treatment, though having longer hydraulic retention time. Tertiary treatment and treatment in a membrane bioreactor (MBR) showed better removal potential than treatment in upflow anaerobic sludge blankets (UASB) in a pilot plant. Human virus genomes were less removed and *Giardia* cysts more removed than all of the studied indicators. Enumeration with PCR, however, may underestimate infectious virion removal. Spores of sulphite-reducing anaerobes and somatic coliphages were significantly less removed than E. *coli* and enterococci in all the studied processes. Bacterial indicator and spore removals correlated to enterovirus genome removal ($p < 0.05$), but the predictive values were low ($R < 0.4$). Removals between microbial indicators and NH_4-N, Kjeldal-N, COD and TOC correlated stronger ($10^{-18} < p < 0.02$; $0.43 < R < 0.90$).

To manage the risk with reuse and discharge of wastewater, treatment performance targets have been calculated as a step in a hazard analysis and critical control point (HACCP) approach. These targets varied from 0 to 10.4 log removal due to water (grey or wastewater), organism (rotavirus, *Campylobacter* or parasitic (oo)cysts) and exposure (drinking water, surface water, aerosols, irrigation of crops or public parks). Faecal contamination in greywater was measured by coprostanol and was shown to be 980 times lower than in wastewater, corresponding to 2.9 log removal in treatment. Somatic coliphages were suggested to function as an index of virus removal in wastewater treatment processes as well as to be included in the monitoring of bathing water. The guideline level was suggested to be 300 PFU 100 mL-1 based on MRA of enteroviruses. This level in a water sample would equal a probability of infection of 0.3% (95th percentile 4%). The risk is overestimated if animal sources dominate the faecal pollution. Development in methods to track sources of faecal pollution showed that if somatic coliphages are enumerated together with phages infecting *Bacteroides* strain GA17, discriminating human from animal faecal pollution is possible based on the ratio between the phages.

Phosphorus release and recovery from treated sewage sludge

Kristina Stark, Architecture and the Built Environment, Royal Institute of Technology (KTH), Stockholm. Defended 14 September 2005.

In working towards a sustainable society, recycling and recovery of products together with handling of scarce resources must be considered. The growing quantities of sludge from wastewater treatment plants and the increasingly stringent restrictions on landfilling and on agricultural use of sludge are promoting other disposal alternatives. Sludge fractionation, providing sludge volume reduction, product recovery and separation of toxic substances into a small stream, has gained particular interest. In this thesis, the potential for phosphate release and recovery from treated sewage sludge is investigated as an alternative for agricultural use in urban areas. Leaching and recovery experiments were performed on sludge residue from supercritical water oxidation, ash from incineration and dried sludge at different temperatures.

Results showed that acid or alkaline leaching is a promising method to release phosphate from sewage sludge treated with supercritical water oxidation, incineration, or drying at 300°C. The leaching is affected by a number of factors, including how the sludge residue has been produced, the origin of the sludge residue, the quantity of chemicals added, and the presence of ions in the leachate.

The implementation of any particular sludge treatment technology would depend on cost, environmental regulations, and social aspects. The results of this thesis may be beneficial for minimizing the use and cost of chemicals, and give increased knowledge for further development of technology for phosphate recovery.

Modelling and control of bilinear systems. Application to the activated sludge process

Mats Ekman, Dept. of Information Technology, Uppsala University, Uppsala, Sweden. Uppsala Dissertations from the Faculty of Science and Technology 65. ISBN 1104-2516. Distributor: Uppsala University Library, Box 510, SE-751 20 Uppsala, Sweden. Defended 21 October 2005.

This thesis concerns modelling and control of bilinear systems (BLS). BLS are linear but not jointly linear in state and control. In the first part of the thesis, a background to BLS and their applications to modelling and control is given. The second part, and likewise the principal theme of this thesis, is dedicated to theoretical aspects of identification, modelling, and control of mainly BLS, but also linear systems. In the last part of the thesis, applications of bilinear and linear modelling and control to the activated sludge process (ASP) are given.

In the system identification part of the thesis special emphasis is devoted to errors-in-variables (EIV) problems for linear as well as bilinear systems. The parameter estimation problem for continuous-time BLS is also investigated. One main point is that both the EIV problem and the continuous-time BLS parameter estimation problem can be treated as separable least-squares (LS) problems. A new bias-eliminating approach, based on a compensated LS solution of an overdetermined system of equations and separable LS, is introduced for identification of parameters in dynamic linear and bilinear systems with EIV, and for parameter estimation of continuous-time BLS.

Two different strategies for controlling BLS are investigated. Firstly, a sub optimal control law for the continuous-time BLS is presented.

Secondly, a model predictive control (MPC) algorithm for discrete-time BLS is presented. The last part of the thesis is focused on applications to wastewater treatment plants. Reduced order time varying bilinear state-space models for the ASP are derived and used for control applications. An adaptive control strategy for control of the nitrate level in an ASP is also presented. Finally, a supervisory aeration volume control strategy for an ASP is discussed. The control strategy is evaluated in a simulation study as well as in a real pilot plant.

Biofilms in Drinking Water Treatment Biofiltration: Membrane Fouling and Regrowth Potential

Frank Persson, CMB, Microbiology, Göteborg University, Göteborg. Defended 9 December 2005.

Drinking water treatment is needed to deliver water at the tap that is safe to drink and has a pleasant look, taste and odour. To obtain this, surface water is commonly treated by chemical coagulation, sedimentation and rapid media filtration (referred to as chemical treatment), prior to disinfection. The treatment processes aim at removing particles, microorganisms, substances that cause unpleasant taste and odour, natural organic matter (NOM) and nutrients (organic and inorganic). The latter fraction is important to prevent microorganisms to proliferate in the distribution systems.

In this work, biofiltration of untreated surface water was investigated. The process was intended as pre-treatment for subsequent chemical treatment or membrane filtration. Granular activated carbon (GAC) and crushed expanded clay (EC) were compared as biofilter media.

GAC and EC of similar grain size were equivalent as filter media for biofiltration regarding development and distribution of microbial biomass. Similar amounts of surface water particles and autofluorescent microalgae of size 0.4-1 μm and 1-15 μm were entrapped in the GAC and EC biofilters with a removal in the range of 60-90%. Microspheres of bacterial size (1 μm) added at high concentrations revealed a high initial removal followed by slow detachment, indicative of a role of the biofilters to level out concentrations of particles and microorganism for subsequent treatment steps. Hydrophobic microspheres were better removed that hydrophilic ones, showing the importance of surface properties of particles and microorganisms, while removal was similar on GAC and EC. The removal of bacteriophages (virus) was limited. Biofiltration with GAC and EC reduced biodegradable organic matter (BOM) by about 30% with a concomitant reduction of biofilm formation in the biofiltered water regarding biofilm total bacterial numbers (60-95%), biovolume (80-90%) and biofilm plate count bacteria (99%).

Chemical treatment showed a removal of particles by 1-2 log over the entire treatment train (1 log=90%, 2 log=99%). Removal of added bacteriophages was considerably higher, at 4-6 log and removal was highly dependent on the initial phase of the coagulation. The removal of particles, autofluorescent microalgae and NOM achieved by the biofilters carried trough the chemical treatment train resulting in an improved total removal of these parameters.

Geosmin and MIB, two substances that cause unpleasant taste and odour, were not reliably removed by the chemical treatment. Biofiltration reduced geosmin and MIB by 80-98%. The biofilter with GAC removed geosmin and

MIB better than the biofilter with EC during unfavourable conditions (low temperature and low biofilter biomass). A difference in the removal mechanisms was observed as geosmin and MIB were removed entirely due to biodegradation on EC, while adsorption played a major role for the removal on GAC despite long operational times.

The combination of biofiltration and nanofiltration (NF-Bio) resulted in significantly less fouling problems than nanofiltration with rapid media filtration pre-treatment (NF-RMF), a process combination that is used at several full-scale nanofiltration plants. The fouling components investigated showed less biological, organic and inorganic fouling on the NF-Bio compared to the NF-RMF membranes. Since biopolymers were the dominant organic fraction, biological fouling was considered important together with iron, aluminium and silica, which were the dominant inorganic elements. The nanofilters removed bacteriophages added to the feed water by 8 log showing that the process was an excellent barrier for microorganisms. NOM was also reduced at a high degree, with only small concentrations of low molecular weight NOM components and BOM in the permeates. Despite this, biofilms formed at a high degree in the nanofiltered water indicative of possibilities for regrowth even at concentrations of BOM below values proposed for biostability.

Phosphorus removal from wastewater with microalgae

Karin Larsdotter, Department of Biotechnology, Royal Institute of Technology (KTH). To be defended in 2006.

A wastewater treatment step with microalgae has been constructed to remove residual phosphorus in the effluent from a hydroponic wastewater treatment pilot plant. The treatment plant is built in a greenhouse and uses a combination of microbiological processes and so called hydroponics, which means that green plants are grown directly in the wastewater itself. The green biomass is also supposed to be harvested and used for different purposes.

Studies on the performance of the microalgal treatment step have been conducted during summer, to see the maximum potential, as well as during the whole year to see the variation in performance as a result of season. The mechanisms for phosphorus removal have also been more thoroughly investigated in lab-experiments.

The results have shown that the treatment method has a great potential during summer, when around 80-95% of the phosphorus can be removed. Also during winter occasions with good phosphorus removal was experienced, but generally the treatment efficiency was low during the darkest months.

The phosphorus removal has been shown to be a combination of biological assimilation by the algae and chemical precipitation of calcium phosphates as a result of the elevated pH levels mediated by the growing algae. This chemical stripping comprises a significant part of the removal, but is still depending on healthy algal growth, and consequently on good light conditions.

The overall evaluation of the algal treatment process suggests algae to be a good way of removing phosphorus from wastewater, even at higher latitudes. In order to increase the removal efficiency over the year, however, improvements regarding bioreactor design can be made. Reactors with short light path, e.g. shallow tanks or transparent tubes, could most probably increase the overall performance since the process was shown to be light limited, even during summer. However, in order to reach effluent levels of phosphorus over the whole year, other biological or chemical phosphorus removal processes may also be needed to supplement the microalgae during winter.

Modelling of Substance Flows in Urban Drainage Systems

Stefan Ahlman, Dept. of Civil and Environmental Technology, Chalmers University of Technology, Göteborg.
To be defended in 2006.

Stormwater is known to be a large contributor of toxic substances to receiving waters. Therefore different measures to manage stormwater quality have been proposed, both structural and non-structural best management practices (BMPs). Today computer models are widely used in the analysis, evaluation and selection of BMPs. To be able to model scenarios related to quality source control, i.e. reducing the pollutants at the source, it is necessary to incorporate a detailed pollutant generation and transport description in such a modelling framework. The sources for pollutants from different activities in the urban area have to be separated in their respective origin, i.e. material corrosion, brake wear, tyre wear etc. The main objective of the study is to develop a modelling framework that enables a source based pollution analysis of urban drainage systems.

The model is called SEWSYS and was developed in MATLAB/Simulink environment. SEWSYS is used for simulating substance flows in urban drainage systems. At present the model includes 21 different substances, including nutrients, heavy metals and organic pollutants. The model framework handles both domestic wastewater and stormwater, in either combined or separate sewer systems. In the stormwater quality module, the pollutants from sources like dry deposition, traffic and construction materials are generated and accumulated during dry weather until they are washed away during rainfall.

Validation studies of the SEWSYS model were carried out using stormwater quality data from Sweden and Australia. Calibration and validation were performed using a split-sample technique. The hydrological part of the model performed well in the validation studies but the quality part was harder to calibrate. This was found to partly depend on the over-parameterisation of the model.

Uncertainty analysis of the model was carried out using a method known as "Preliminary Uncertainty Analysis". This method included a systematic unravelling of the model's components into model structure, attributes, input values, model parameters, state variables and output values. The statistical uncertainty for each quantity was assessed using a priori knowledge and literature values, followed by Monte Carlo simulations to calculate the total output uncertainty.

Based on the level of detail in the model's output for stormwater pollution, it is clear that the model represents a useful tool for simulating and evaluating quality source control measures.

Integrating farming and wastewater management– a systems perspective

Pernilla Tidåker, Dept. of Biometry and Engineering, Swedish University of Agricultural Sciences, Uppsala, Sweden.
To be defended in 2006.

In a sustainable society, the plant nutrients in sewage products have to be recycled in a sustainable way. Sewage sludge is the predominant sewage fertiliser product available today, but the use of sewage sludge in agriculture has long been questioned as regards hazardous substances. In this respect, source-separated products better seem to fulfil agricultural requirements regarding fertiliser products. By replacing mineral fertiliser products, emissions and resource use related to production of those are avoided. However, new wastewater systems require infrastructural changes that also are associated with environmental burdens. These changes might concern storage, sanitation, transport and spreading of the sewage products as well as impacts on the already existing conventional treatment. This doctoral project highlighted different aspects on systems integrating farming and wastewater management. The main objective was to analyse resource use and environmental impacts when sewage products replaced mineral fertilisers in agricultural production and identify critical factors for a beneficial management of sewage products that will promote system sustainability.

Three scenario studies were performed within the doctoral project. The methodology used was based on a life cycle perspective, including relevant aspects from raw material acquisition through production, use and disposal. In the first study, conventional wheat production was compared with a scenario where source-separated human urine replaced mineral fertilisers. The second study modelled the environmental consequences when a country town introduced a system based on separate handling of the blackwater that after sanitation served as fertiliser. The third study focused on different options based on recycling principles for improving current on-site systems in Sweden.

The largest relative impact compared to the total anthropogenic impacts from Sweden was in general the reduced eutrophying emissions to water and the replaced amount of phosphorus fertilisers. An optimal fertilising strategy regarding application time and technique, and achieving a high substitution of mineral fertiliser were demonstrated to be important for many environmental and resource aspects. For reducing the energy use, a well designed collection system was shown to be important. The choice of method for sanitising the sewage products was another crucial topic. Urea treatment proved a promising alternative for small-scale systems as urea will serve as fertiliser at a later stage.

Applying an agricultural perspective when evaluating systems integrating wastewater management and agriculture also highlighted potential conflicts regarding nutrient utilisation.

10.3 REFERENCES

Ahlman, S. (2000) *SEWSYS - ett modelleringsverktyg för transport - och reningsprocesser i avloppssystem utvecklat i MATLAB-Simulink* (SEWSYS - a modelling tool for transport and treatment processes in sewer systems developed in MATLAB-Simulink, in Swedish). M.Sc. Thesis 2000:8, Water Environment Transport, Chalmers University of Technology, Göteborg, Sweden.

Ahlman, S. and Svensson, G. (2002) Modelling substance flows in urban sewer systems using MATLAB/Simulink. In: Strecker, E.W. and Huber, W.C. (eds.) *Proceedings of the 9th International Conference on Urban Drainage*, September 8-13 2002, Portland, OR, USA.

Ahlman, S., Svensson, G., Malm, A., Kant, H. and Karlsson, P. (2004a) Methodology for systems analysis of sustainable wastewater management - assessment for an urban catchment in Vasastaden, Göteborg. In: *Proceedings of the 5th International Conference on Sustainable Techniques and Strategies in Urban Water Management (NOVATECH'2004)*, June 6-10 2004, Lyon, France.

Ahlman, S., Kant, H., Karlsson, P., Malm, A. and Svensson, G. (2004b) *Systemanalys Vasastaden i Göteborg – Avloppsvattensystemet* (System Analysis in Vasastaden, Göteborg – The wastewater system, in Swedish). Urban Water report 2004:5, Chalmers University of Technology, Göteborg. pdf: www.urbanwater.org

Ahlman, S. and Svensson, G. (2005) *SEWSYS– a tool for simulation of substance flows in urban sewer systems.* Urban Water report 2005:11, Chalmers University of Technology, Göteborg. pdf: www.urbanwater.org

Ahlman, S., Malm, A., Kant, H., Svensson, G. and Karlsson, P. (2005a) Modelling non-structural Best Management Practices – focus on reductions in stormwater pollution. *Water Science and Technology* **52**(5), 9-16.

Ahlman, S., Svensson, G., Cziemel Berntsson, J. and Bengtsson, L. (2005b) *Alternativa lösningar för dagvattenhantering – en delrapport inom studien uthålliga VA-system i Uppsala* (Alternative solutions for stormwater management – a report from the model city studies of sustainable urban water management in Uppsala, in Swedish), Urban Water report 2005:12, Chalmers University of Technology, Göteborg. pdf: www.urbanwater.org

Akiba, M., Kunikane, S., Kim, H.S. and Kitazawa, H. (2002) Algae as surrogate indices for the removal of Cryptosporidium oocysts by direct filtration. *Water Science and Technology: Water Supply*, **2**(3), 73-80.

Albrechtsen, H.J. (2002) Microbiological investigations of rainwater and greywater collected for toilet flushing. *Water Science and Technology*, 46 (6-7), 311-316.

Alegre, H., Hirner, W., Baptista, J.M. and Parena, R. (2000) *Performance indicators for water supply services.* Manual of Best Practice Series, IWA Publishing, London, UK. ISBN: 1900222272.

Alegre, H. and Baptista, J.M. (2003) WP1 – *Construction of a control panel of performance indicators for rehabilitation: Rehab PI guidance range.* Report No. 1.5 of the project CARE-W: Computer Aided Rehabilitation of Water networks. 5th Framework Programme of the European Union. LNEC (Laboratorio Nacional de Engenharia Civil), Lisbon, Portugal.
electronic edition: http://www2.unife.it/bscw/bscw.cgi/.

Anderson, W.B., Slawson, R.M. and Mayfield, C.I. (2002) A review of drinking-water-associated endotoxin, including potential routes of human exposure. *Canadian Journal of Microbiology*, **48**(7), 567-587.

Andersson, Å. and Jenssen A. (2002) *Flöden och sammansättning på BDT-vatten, urin, fekalier och fast organiskt avfall i Gebers*; (Flows and composition of greywater, urine, faeces and solid biodegradable waste in Gebers; in Swedish with English abstract). Institutionsmeddelande (Report) 2002:05, Department of Agricultural Engineering, Swedish University of Agricultural Sciences. Pdf: www.slu.se.

Andrén, S., Arderup, M. and Hornborg, A. (ed.) (2004) *Humanekologiska perspektiv på hållbar produktion och konsumtion.* (An integrated perspective on sustainable production and consumption, in Swedish). Rapport 5354, Naturvårdsverket (Report, Swedish Environmental Protection Agency), Stockholm. pdf: www.naturvardsverket.se

Anonymous (2001) Waterleidingbesluit 2001 (In Dutch). *Staatsblad van het Koninkrijk der Nederlanden* **31**, 1-53.

Argyris, C. and Schön, D. (1996) *Organizational Learning II: Theory, Method, and Practice.* Addison-Wesley, Reading, Massachusetts, USA.

Ashbolt, N.J., Grabow, W.O.K. and Snozzi, M. (2001) Indicators of microbial water quality. In: Fewtrell, L. and Bartram, J. (eds.) *Water Quality: Guidelines, Standards and Health. Risk assessment and management for water-related infectious disease,* Chapter 13, pp. 289-315. IWA Publishing, London, UK.

Ashbolt, N.J., Petterson, S.R., Roser, D.J., Westrell, T., Ottoson, J., Schönning, C. and Stenström, T.A. (2004) Microbial risk assessment tool to aid in the selection of sustainable urban water systems, *Proceedings of the 2nd IWA Leading-Edge Conference on Sustainability: Sustainability in Water Limited Environments - LES2004,* 8-10 November 2004, Sydney. IWA Publishing, London.

Ashbolt, N.J., Petterson, S.R., Stenström, T.A., Schönning, C., Westrell T. and Ottoson, J. (2005) *Microbial Risk Assessment (MRA) Tool.* Urban Water report 2005:7. Chalmers University of Technology, Göteborg, Sweden. pdf: www.urbanwater.org

Ashley, R., Blackwood, D., Butler, D. and Jowitt, P. (eds.). (2004) *Sustainable Water Services - A procedural guide,* IWA Publishing, London.

Axelsson, K., Delefors, C. and Söderström, P. (2001) *Hammarby Sjöstad – en kvalitativ studie av människors faktiska miljöbeteende och dess orsaker.* (Hammarby Sjöstad – a qualitative study on peoples' actual environmental behaviour and its causes, in Swedish) Rapport nr. 26, Stockholm Vatten (Report, Stockholm Water Co.), Sweden.

Azar, C., Holmberg, J. and Karlsson, S. (2002) *Decoupling – past trends and prospects for the future.* Environmental Advisory Council, Swedish Ministry of the Environment, Stockholm. ISSN: 0375-250X. pdf: www.mvb.gov.se

Bartram J., Fewtrell L. and Stenström T.A. (2001) Harmonised assessment of risk and risk management for water-related infectious disease. An overview. In: Fewtrell, L. and Bartram, J. (eds.) *Water Quality – Guidelines, Standards and Health: Assessment of Risk and Risk Management for Water-Related Infectious Disease.* World Health Organization and IWA.

Baumol, W. and Oates, W. (1988) *The Theory of Environmental Policy.* Cambridge University Press, Bristol, UK.

Baun A., Eriksson E., Ledin A. and Mikkelsen P.S. (2005) Screening tool for problem and hazard identification of xenobiotic organic compounds in stormwater. (Submitted for publication).

Beierle, T.C., and Cayford, J. (2002) *Democracy in practice. Public participation in environmental decisions.* RFF Press, Washington DC.

Bergstedt, O. and Rydberg, H. (2002) Naturally occurring autofluorescent particles as surrogate indicator of sub-optimal pathogen removal in drinking water. In: Hahn.

H.H., Hoffmann, E., Ødegaard, H. (eds.) *Chemical Water and Wastewater Treatment*, vol. VII., IWA Publishing, London, 191–200.

Björklund, A. (1998) *Environmental systems analysis of waste management with emphasis on substance flows and environmental impact*. Licentiate Thesis, Department of Chemical Engineering and Technology, Royal Institute of Technology, Stockholm, Sweden.

Björlenius, B. (2003) Lokalt reningsverk för Hammarby Sjöstad – en plattform för jämförande studier av aeroba och anaeroba processer samt membranteknik (A local wastewater treatment plant for Hammarby Sjöstad – a possibility to studies comparing aerobic, anaerobic and membrane processes, in Swedish), *Proceedings of the "8:e Nordiska Avloppsvattenkonferensen"*, 17-19 November.Esbo, Finland.

Blomqvist, A. (2002) *Can Watercourse Groups Reduce Nutrient Losses?* VASTRA Report no. 4, Sweden. pdf: www.vastra.org.

Bonn Charter (2004) *The Bonn Charter for Safe Drinking Water*. International Water Association, London, UK.

Bossel, H. (1997) Finding a Comprehensive Set of Indicators of Sustainable Development by Application of Orientation Theory. In: Moldan, B., Billharz, S., and Matraves, R. (eds.): *Sustainability Indicators: A Report on the Project on Indicators of Sustainable Development*. John Wiley & Sons, New York.

Bossel, H. (1998) *Earth at a Crossroad. Paths to a Sustainable Future*. Cambridge University Press, Bristol, UK.

Brouwer, R. (2000) Environmental value transfer: State of the art and future prospects. *Ecological economics* **32**, 137-152.

Brundick, B. (ed.) (1995) *Parametric Cost Estimating Handbook*. NASA. pdf: www.jsc.nasa.gov/bu2

Butler, D., Jowitt, P., Ashley, R., Blackwood, D., Davies, J., Oltean-Dumbrava, C., McIlkenny, G., Foxon, T, Gilmour, D., Smith, H., Cavill, S., Leach, M., Pearson, P., Gouda, H., Samson, W., Souter, M., Hendry, S., Moir, J. and Bouchart, F. (2003) Sward - Decision support processes for the UK water industry. *Management of Environmental Quality* **14**(4), 444-459.

Byrns, G. (2001) The fate of xenobiotic organic compounds in wastewater treatment plants. *Water Research* **35**(10), 2523-2533.

Carlsson Reich, M. (2002) *Samhällsekonomisk analys av system för återanvändning av fosfor ur avlopp* (National economic analysis of a system to recycle phosphorous from sewage, in Swedish with English summary). Report no. 5222. Swedish Environmental Protection Agency, Stockholm. pdf: www.internat.naturvardsverket.se

Chorus, I. and Bartram, J. (eds.) (1999) *Toxic cyanobacteria in water: A guide to their public health consequences, monitoring and management*. E & FN Spon Publ., London.

Costanza, R. (ed.) (1991) *The Science and Management of Sustainability*. Columbia University Press, New York.

Côté, P. and Thompson, D. (2000) Wastewater treatment using membranes: the North American experience. *Water Science and Technology* **41**(10-11), 209-215.

Crettaz, P., Jolliet, O., Cuanillon, J.M. and Orlando, S. (1999) Life cycle assessment of drinking water and rain water for toilets flushing. *J. Wat. Suppl.: Res. & Techn. – AQUA* **48**(3), 73-83.

Dalemo, M. (1999) *Environmental Systems Analysis of Organic Waste Management – the ORWARE Model and the Sewage Plant and Anaerobic Digestion Submodels*. Ph.D Thesis, Dept of Agricultural Engineering, Swedish University of Agricultural Sciences, Uppsala, Sweden.

Daughton, C.G. and Ternes, T.A. (1999) Pharmaceuticals and personal care products in the Environment: Agents of subtle Change? *Environmental Health Perspectives*, **107**(6), 907-938.

Drangert, J.-O. and Krantz, H. (2002) *Hammarby Sjöstad – miljöföreställningar och verkligheter.* (Hammarby Sjöstad – residents' environmental perceptions and realities, in Swedish) Vatten **58**, 89-95.

Drangert, J.-O. (2004) Managing resource flows in a sustainable city. In: *Abstract Volume of the 14th Stockholm Water Symposium,* August 16-20, 2004. Stockholm International Water Institute. pp. 372-3.

Drangert, J.-O. and Löwgren, M. (2005) *Förändring eller kontinuitet? Faktorer som påverkar va-systemens utveckling i Linköping och Norrköping under perioden 1960 – 1990.* (Change or Continuity? Factors impacting the evolution of the water and sanitation systems in Linköping and Norrköping cities in the period 1960 – 1990, in Swedish) Urban Water report 2005:1, Chalmers University of Technology, Göteborg. pdf: www.urbanwater.org

Drangert, J.-O., Klockner, A. and Nors, L. (2005) *På väg mot en hållbar stad – uppfattad och uppmätt påverkan av miljösatsningar i Hammarby Sjöstad, Stockholm.* (Towards a sustainable city – perceived and quantified influence from environmental concentrations in Hammarby Sjöstad, Stockholm, in Swedish) VA-Forsk rapport 2005-08, Svenskt Vatten (Report, Swedish Water and Wastewater Association), Stockholm. pdf: www.svenkstvatten.se

EEA (1999) *Guidelines for defining and documenting data on costs of possible environmental protection measures.* Technical Report 27, European Environmental Agency.

Ekins, P. (2000) *Economic Growth and Environmental Sustainability. The Prospects for Green Growth.* Routhledge, London, UK.

Ekman, M. (2005) *Innovative information systems.* Ph.D. Thesis. Dept. of Information Technology, Uppsala University, Uppsala, Sweden. pdf: www.urbanwater.org

Eliasson, G., Tykesson, E., Jansen, J. La C. and Hansen, B. (2000) Utilisation of fractions of digester sludge after thermal hydrolysis. In: Hahn, H.H., Hoffmann, E. and Ødegaard, H. (eds.) *Chemical Water and Wastewater Treatment VI.* Springer Verlag, Heidelberg, pp. 339–345.

Emelko, M.B. (2003) Removal of viable and inactivated Cryptosporidium by dual- and tri-media filtration. *Wat. Res.* **37**(12), 2998-3008.

Engvall, C. (1999) *Simulation of Materials Flow in Stormwater.* M.Sc. Thesis, Uppsala University School of Engineering, Uppsala, Sweden.

Ericsson, N., Vinnerås, B. and Jönsson, H. (2005) *Källsorterande toaletter – brukarnas erfarenheter, problem och lösningar* (Source separating toilets – User experience, problems and solutions; in Swedish, English abstract). Rapport – miljö, teknik och lantbruk (Report) 2005:01, Department of Biometry and Engineering, Swedish University of Agricultural Sciences, Uppsala. Pdf: www.slu.se.

Eriksson, E., Auffarth, K, Henze, M., and Ledin, A. (2002) Characteristics of grey wastewater. *Urban Water* **4**, 85-104.

Eriksson, E., Auffarth, K., Eilersen, A.M., Henze, M., and Ledin, A. (2003) Household chemicals and personal care products as sources for xenobiotic organic compounds in grey wastewater. *Water S.A.* **29**(2), 135-146.

Eriksson, E., Baun, A., Mikkelsen, P.S. and Ledin, A. (2005) Chemical hazard identification and assessment tool for evaluation of stormwater priority pollutants. *Water Science and Technology* **51**(2), 47-55.

Eriksson E., Baun A., Mikkelsen P.S., and Ledin A. (2006) Characteristics and quality of urban runoff – a review. (manuscript in preparation).

Eriksson, O., Frostell, B., Björklund, A., Assefa, G., Sundqvist, J.-O., Granath, J., Carlsson, M., Baky, A. and Thyselius, L. (2002) ORWARE – a simulation tool for waste management. Resources, *Conservation and Recycling* **36**(4), 287-307.

Etnier, C. and Söderberg, H. (2002) Multicriteria assessment for choosing a wastewater treatment option: the case of Hammarby Sjöstad, Sweden. *Proceedings National Onsite Wastewater Recycling Association Conference*, Kansas City, USA, September 2002.

EU REACH Programme (2005) Access date: 2005-05-04. http://europa.eu.int/comm/environment/chemicals/reach.htm

EU Water Framework Directive (2000) *Directive 2000/60/EC of the European Parliament and of the Council of 23 October 2000 establishing a framework for Community action in the field of water policy.* http://europa.eu.int/

European Community (2001a) *White Paper - Strategy for a future chemicals policy.* Brussels, Belgium, February 27, 2001.

European Commission (2001b) *On the Sixth Environmental Action Programme of the European Community, Environment 2010: Our Future, Our Choice.* Brussels, Belgium, pdf: http://europa.eu.int/eurlex/en/com/pdf/2001/en_501PC0031.pdf.

European Commission (2003) *Technical Guidance Document in support of Commission Directive 93/67/EEC on Risk Assessment for new notified substances, Commission Regulation (EC) No 1488/94 on Risk Assessment for existing substances and Directive 98/8/EC of the European Parliament and of the Council concerning the placing of biocidal products on the market.* Brussels, Belgium.

European Commission (2004) *Establishing the list of priority substances in the field of water policy and amending Directive 2000/60/EC.* pdf: http://europa.eu.int/eurlex/pri/en/oj/dat/2001/l_331/l_33120011215en00010005.pdf

Fane, S. and Ashbolt, N.J. (2000) A methodology for assessing comparative pathogen risks of novel systems recycling waters and nutrients. In: Dillon, P.J. (ed.) *Water Recycling Australia.* pp. 11-16. CSIRO Land and Water, Australian Water Association, Adelaide.

Friend, J. and Hickling, A. (2005) Planning Under Pressure: *The Strategic Choice Approach* (3rd ed.). Elsevier Architectural Press, Oxford, UK.

Gannholm C. (2004) *Utvärdering av fjärravlästa individuella vattenmätare i Hammarby Sjöstad – Tillförlitlighet och miljömål* (Evaluation of remote-read individual water meters in Hammarby Sjöstad – reliablity and environmental goals, in Swedish). Rapport nr. 37-2004. Stockholm Vatten (Report, Stockholm Water Co.), Sweden.

Gidner A., Almemark M., Stenmark L. and Östengren, Ö. (2000) Treatment of sewage sludge by supercritical water oxidation. *IBC's 6th Annual Conference on Sludge.* 16-17 Feb. 2000, London, UK.

Grimvall, G., Jacobsson, P. and Thedéen, T. (2003) *Risker i tekniska system* (Risk in technical systems; Course book, in Swedish) Studentlitteratur, Lund, Sweden.

Haas, C.N., Rose, J.B. and Gerba, C.P. (1999) *Quantitative Microbial Risk Assessment.* John Wiley & Sons Inc., New York.

Hanley, N. and Spash, C. (1993) *Cost-benefit analysis and the Environment.* Edward Elgar Publishing Ltd, Cheltenham, UK.

Hardi, P. and Barg, S. (1997) *Measuring sustainable development: Review of current practice.* Occasional paper number 17, Industry Canada, Ottawa, Canada.

Healey, P. (1997) *Collaborative Planning: Shaping Places in Fragmented Societies.* Macmillan Press Ltd., Hong Kong.

Heberer, T. (2002) Occurrence, fate and removal of pharmaceutical residues in the aquatic environment: a review of recent research data. *Toxicology Letters* **131**, 5-17.

Heinicke, G., Långmark, J., Persson, F., Hedberg, T. and Storey, M.V. (2004) Significance of dosage point for challenge tests in coagulation treatment. In: Hahn, H., Hoffmann, H. and Ødegaard, H. (eds.) *Chemical Water and Wastewater Treatment*, vol. VIII. 191-200. IWA Publishing, London. ISBN: 1-84339-068-X

Heinicke, G. (2005) *Biological Pre-filtration and Surface Water Treatment – Microbial barrier function and removal of natural inorganic and organic compounds.* Ph.D. Thesis, Water Environment Transport, Chalmers University of Technology, Göteborg. Available: www.urbanwater.org

Heinicke, G., Persson, F., Uhl, W., Hermansson, M. and Hedberg, T. (2005) The effect of biological pre-filtration on the performance of conventional surface water treatment. *J. Wat. Suppl.: Res. & Techn. – AQUA, in press.*

Hellström, D., Jeppsson, U. and Kärrman, E. (2000) A framework for systems analysis of sustainable urban water management. *Environmental Impact Assessment Review* **20**(3), 311-321.

Hellström, D., Baky, A., Palm O., Jeppsson, U. and Palmquist, H. (2003) Comparison of Resource Efficiency of Systems for Management of Toilet Waste and Organic Household Waste, *Proceedings of the 2nd Int. Sympos. Ecological Sanitation*, April 7 – 11, 2003, Lübeck, Germany.

Hellström, D., Hjerpe, M. and Van Moeffaert, D. (2004a) *Indicators to assess ecological sustainability in the Urban Water Sector.* Urban Water report 2004:3, Chalmers University of Technology, Göteborg. pdf: www.urbanwater.org

Hellström, D., Kärrman, E., Rydhagen, B. and Palm, O., (2004b), Svartvattensystem i Hammarby Sjöstad? – Systemanalys och förslag på utformning av system (Blackwater treatment in Hammarby Sjöstad? Systems analysis and design recommendations, in Swedish), *Vatten* **60**(3), 201-208.

Hellström, D. (ed.) (2005) *Slutrapport från modellstaden Hammarby Sjöstad* (Final report on the model city Hammarby Sjöstad, in Swedish). Urban Water report 2005:4. Chalmers University of Technology, Göteborg, Sweden. pdf: www.urbanwater.org

Hellström, D., Hjerpe, M. and Van Moeffaert, D. (2005) *Indicators to assess ecological sustainability in the Urban Water sector.* Urban Water report 2004:3. Chalmers University of Technology, Göteborg, Sweden. pdf: www.urbanwater.org

Henze, M., Gujer, W., Mino, T. and van Loosdrecht, M. (2000) *Activated Sludge Models ASM1, ASM2, ASM2d and ASM3.* Scientific and Technical Report No 9., IWA Publishing, London, UK.

Hiessl, H. and Toussaint, D. (2003) Options for sustainable urban water infrastructure systems: Results of the AKWA 2100 project. *Proc. 2nd Ecosan International Symposium of Ecological Sanitation*, Lübeck, Germany.

Hiessl, H., Toussaint, D., Becker, M., Dyrbusch, A., Geisler, S., Herbst, H. and Prager, J.U. (eds.) (2003) *Alternativen der kommunalen Wasserversorgung und Abwasserentsorgung AKWA 2100* (Alternatives for the municipal water supply and wastewater handling, in German). Schriftenreihe des Fraunhofer-Instituts für Systemtechnik und Innovationsforschung ISI - Technik, Wirtschaft und Politik, Physica-Verlag, Heidelberg,

Hjerpe, M. and Krantz, H. (2001) Handling of stormwater in open structures: a case study of critical factors for sustainability. *Proceedings from the 2nd IWA 2001 World Water Congress*, Berlin, Germany, 15-19 October 2001, CD-ROM.

Hjerpe, M. and Löwgren, M. (2002) Identifying economic indicators of sustainable urban water management. *Poster at the IWA Leading Edge conference: Sustainabltility in the water sector*, Venice, Italy, 25-26 November 2002.

Hjerpe, M. (2005) *Sustainable development and urban water management: Linking theory and practice of economic criteria.* Ph.D. Thesis, Linköping Studies in Arts and Science, no. 322, Linköping University, Linköping, Sweden. pdf: www.urbanwater.org

Höglund, C., Ashbolt, N.J., Stenström, T.A. and Svensson, L. (2002a) Viral persistence in source-separated human urine. *Advances in Environmental Research* 6(3), 265-275.

Höglund, C., Stenström, T.A., Ashbolt, N.J. (2002b) Microbial risk assessment of source-separated urine used in agriculture. *Waste Management and Research* 20(3), 150-161.

Hurst, A.M., Edwards, M.J., Chipps, M., Jefferson, B. and Parsons, S.A. (2004) The impact of rainstorm events on coagulation and clarifier performance in potable water treatment. *Science of the Total Environment* 321(1-3), 219-230.

Imboden, D. (2000) Energy forecasting and atmospheric CO_2 perspectives: two worlds ignore each other. *Integrated Assessment* 1(4), 321-330.

Jansen, J.L.C., Gruvberger, C., Hanner, N., Aspegren, H. and Svärd, A. (2004) Digestion of sludge and organic waste in the sustainability concept for Malmo, Sweden. *Water Science And Technology* 49(10): 163-169.

Jeppsson, U. and Hellström, D. (2001) Systems analysis for environmental assessment of urban water and wastewater systems. *Proc. 2nd IWA World Water Congress*, Berlin, Germany, 15-19 Oct. 2001. CD-ROM.

Jeppsson, U., Baky, A., Hellström, D., Jönsson, H. and Kärrman, E. (2005) *The URWARE wastewater treatment plant models.* Urban Water Report 2005:5. Chalmers University of Technology, Göteborg, Sweden. pdf: www.urbanwater.org

Jernlid, A.-S. and Karlsson, K. (1997) *Våta toalettsystem med urinsortering - driftstudier och utvärdering med livscykelanalys* (Water flushed urine-diverting toilet systems – studies of operation and assessment using LCA methodology; in Swedish). M.Sc. Thesis 1997:6, Department of Sanitary Engineering, Chalmers University of Technology, Göteborg, Sweden.

Johansson, A. and Scott, S. (2004) *Den kemiska reningens barriärverkan mot patogena mikroorganismer i dricksvattenberedningen (The microbial barrier function of conventional drinking water treatment,* in Swedish). M.Sc. Thesis 2004:14, Water Environment Transport, Chalmers University of Technology, Göteborg. Available: www.chalmers.se

Johansson, M., Jönsson, H., Höglund, C., Richert Stintzing, A. and Rodhe, L. (2001) *Urine separation – closing the nutrient cycle.* Stockholm Water Company. Stockholm, Sweden. Pdf: www.stockholmvatten.se/pdf_arkiv/english/Urinsep_eng.pdf.

Jönsson, H., Vinnerås, B., Höglund, C., Stenström, T.A., Dalhammar, G. and Kirchmann, H. (2000) *Källsorterad humanurin i kretslopp* (Recycling of source separated human urine, in Swedish). VA-Forsk rapport 2000-1, Svenskt Vatten (Report, Swedish Water and Wastewater Association), Stockholm. pdf: www.svensktvatten.se

Jönsson, H., Baky, A., Jeppsson, U., Hellström, D. and Kärrman, E. (2005a) *Composition of urine, faeces, greywater and bio-waste - for utilisation in the URWARE model.* Urban Water report 2005:6, Chalmers University of Technology, Göteborg. pdf: www.urbanwater.org.

Jönsson, H., Ashbolt, N., Baky, A., Drangert, J.-O., Krantz, H., Kärrman, E., Ledin, A., Ottoson, J., Palmquist, H., Westrell, T. and Vinnerås, B. (2005b) *Slutrapport från modellstaden Urbana enklaven* (Final report for the model city Urban Enclave; in

Swedish, English abstract). Urban Water report 2005:8, Chalmers University of Technology, Göteborg. pdf: www.urbanwater.org

Kain, J-H. (2003) *Sociotechnical Knowledge: An Operationalised Approach to Localised Infrastructure Planning and Sustainable Urban Development*. Ph.D. Thesis, Dept. of Built Environment and Sustainable Development, Chalmers University of Technology, Göteborg, Sweden.

Kain, J.-H. and Söderberg, H. (2005) Management of Complex Knowledge in Planning for Sustainable Urban Development. In: *Proc. International Conference for Integrating Urban Knowledge and Practice*, Göteborg, Sweden. CD-ROM.

Karlsson K., Malmqvist, P.-A., Viklander, M. and Palmquist, H. (2004) The Barrier approach for stormwater systems. *Proceedings of the 5th International conference on innovative technologies in urban storm drainage, NOVATECH 2004*, June 2004 Lyon, France, pp. 591-598.

Kärrman, E., Jönsson, H., Gruvberger, C., Dalemo, M., Sonesson, U. and Stenström, T.A. (1999) *Miljösystemanalys av hushållens avlopp och organiska avfall* (Environmental systems analysis of wastewater and solid organic waste from households, in Swedish). VA-Forsk rapport 1999-15, Svenskt Vatten (Report, Swedish Water and Wastewater Association), Stockholm. pdf: www.svensktvatten.se

Kärrman, E., Söderberg, H., Åberg, H., Olin, B and Van Moeffaert, D. (2005a) *Uthålliga system för avlopp och avfall i Surahammar*. (Sustainable Systems for Wastewater and Solid Organic Waste Management i Surahammar, in Swedish). Urban Water report 2005:10, Chalmers University of Technology, Göteborg. pdf: www.urbanwater.org.

Kärrman, E., Van Moeffaert, D., Johansson, M., Kvarnström, E., Byström, Y., af Petersens, E., Ridderstolpe, P., Olin, B., Baky, A., Hannerz, N., Palm, O., Wittgren, H.-B. and Christensen, J. (2005b) *Avlopp i kretslopp* (Wastewater closing the loop, in Swedish). Rapport (Report) 5406, Swedish Environmental Protection Agency. Pdf: www.naturvardsverket.se.

Kärrman, E., Berg, P., Olin, B., Palme, U., Rydhagen, B., Schönning, C., Söderberg, H. and Tälleklint, M. (2005c) *Uthålliga spillvattensystem i Uppsala*, (Sustainable wastewater management in Uppsala, in Swedish) Urban Water report 2005:9, Chalmers University of Technology, Göteborg. pdf: www.urbanwater.org

Krantz, H. (2005) *Matter that matters. A study of household routines in a process of changing water and sanitation arrangements*. Ph.D. Thesis, Linköping Studies in Arts and Science no 316, Linköping University, Linköping, Sweden. pdf: www.ep.liu.se/diss/arts_science/2005/316/ or www.urbanwater.org

Kroiss, H. (2004) What is the potential for utilizing the resources in sludge? *Water Science and Technology* 49(10), 1-10.

Långmark, J. (2004) *Biofilms and microbial barriers in drinking water treatment and distribution*, Ph.D. Thesis, Dept. of Land and Water Resources Engineering, Royal Institute of Technology (KTH), Stockholm. Available: www.urbanwater.org

Långmark, J., Storey, M.V., Ashbolt N.J. and Stenström, T.A. (2005) Accumulation and fate of microorganisms and micro-spheres in biofilms formed in a pilot-scale water distribution system. *Applied and Environmental Microbiology* 71(2), 706-712.

LeBlanc, R.J., Allain, C.J., Laughton, P.J. and Henry, J.G. (2004) Integrated, long term, sustainable, cost effective biosolids management at a large Canadian wastewater treatment facility. *Water Science and Technology*, 49(10), 155-162.

Ledin A., Auffarth K., Boe-Hansen R., Eriksson E., Albrechtsen H.-J., Baun A. and Mikkelsen P.S. (2004) *Brug af regnvand opsamlet fra tage og befæstede arealer -*

Udpegning af relevante måleparametre. (Use of rainwater collected from non-permeable surfaces Identification of parameters relevant for a monitoring program, in Danish). Miljøstyrelsen, København. Økologisk Byfornyelse og Spildevandsrensning, 48 2004, 1-120. (Report, Danish Environmental Protection Agency). pdf: www.er.dtu.dk/publications/fulltext/2004/MR2002-072.pdf

Ledin A., Aabling T., Baun A., Eriksson E. and Mikkelsen P.S. (2005) *CHIAT - Chemical Hazard Identification and Assessment Tool. En metodik för utvärdering av kemiska risker i samband med hantering av dag- och avloppsvatten* (A methodology for chemical risk assessment of different strategies for handling of storm- and wastewater; in Swedish, English summary). VA-Forsk rapport 2005-9, Svenskt Vatten (Report, Swedish Water and Wastewater Association), Stockholm. pdf: www.svenkstvatten.se

Lodder, W.J., Vinj, J., van de Heide, R., de Roda Husman, A.M., Leenen, E.J.T.M. and Koopmans, M.P.G. (1999) Molecular detection of Norwalk-like caliciviruses in sewage. *Applied and Environmental Microbiology* **65**, 5624-5627.

Loll, A. (1892) *Neues Kanalisations-System nach M.P. von Nadein. (New sewer system according to M.P. von Nadein,* in German) *Gesundheits-Ingenieur* **15**(19), 621-624.

Löwgren, M. (2003) *Den ekonomiska analysen i Vattendirektivet. Ett pilotprojekt i Motala Ströms avrinningsområde* (Economic analysis in the Water Framework Directive. A pilot study in the Motala River Basin, in Swedish with English summary). pdf: www.vastra.org

Ma, W., Sun, Z., Wang, Z., Feng, Y.B., Wang, T C., Chan, U., Miu, C.H. and Zhu, S. (1998) Application of membrane technology for drinking water. *Desalination* **119**, 127-131.

Magnusson, J. (2003) *Characteristics of household wastewater in Hammarby Sjöstad: Hazardous substances and their possible sources.* M.Sc. Thesis 2003:292, Dept. of Environmental Engineering. Luleå University of Technology, Luleå, Sweden.

Malmqvist, P.-A., Ashbolt, N.J., Fane, S., Hellström, D., Jeppsson, U. and Söderberg, H. (2000) Assessing alternative wastewater systems in Hammarby Sjöstad, Stockholm. In: Mangin, J-C. and Miramond, M. (eds.) *Proceedings 2nd Intl. Conference on Decision Making in Urban and Civil Engineering.* pp. 1-13. CUST Clermond-Ferrand, LIP6 Paris and Univ. de Valenciennes, Lyon, France.

Malmqvist, P.-A. and Palmquist, H. (2005) Decision support tools for urban water and wastewater systems - focussing on hazardous flows assessment. *Water Science and Technology* **51**(8), 41-49.

Matos, M.R., Cardoso, M.A., Ashley, R, Duarte, P., Molinari, A. and Shulz, A. (2003a) *Performance Indicators for Wastewaters Services.* Manual of Best Practice Series. IWA Publishing, London. ISBN: 1900222906

Matos, M.R., Cardoso, M.A., Pinheiro, I. and Almeida, M.C. (2003b) *WP1 report - Deliverable D1 - Construction of a control panel of performance indicators for rehabilitation, selection of a listing of rehabilitation PIs.* Project CARE-S – Computer Aided Rehabilitation of Sewer Networks. Decision Support Tools for Sustainable Sewer Network Management, 5th Framework Programme of the European Union, LNEC (Laboratorio Nacional de Engenharia Civil), Lisbon, Portugal.

McCann, B. (2004) Are three qualities better than one? *Water21* (April), p. 35.

McKean, M.A. (1992) Success on the Commons – A Comparative Examination of Institutions for Common Property Resource Management. *Journal of Theoretical Politics* **4**, 247-281.

Mikkelsen, P.S., Adeler, O.F., Albrechtsen, H.J. and Henze, M. (1999) Collected rainfall as a water source in Danish households – What is the potential and what are the costs? *Water Science and Technology* **39**(5), 49-56.

Mikkelsen, P.S., Baun, A. and Ledin, A. (2001) Risk assessment of stormwater contaminants following discharge to soil, groundwater or surface water. In: Marsalek J. (ed.) *Advances in urban stormwater and agricultural runoff source controls*, pp. 69-80. Kluwer Academic Publishers, Dordrecht, NL.

Nordtest (2003) *Increase in colour and amount of organic matter in surface waters.* Position paper 009, Nordic Innovation Centre. Available: www.nordtest.org.

Okun, D.A. (2000) Water reclamation and unrestricted nonpotable reuse: A new tool in urban water management. *Annual Review of Public Health* **21**, 223-245.

Olin, B., Kant, H. and Ramirez, J.-I. (2005) *Kostnadsmodell för strategiska vägval* (Costing model for strategic choices, in Swedish). VA-Forsk rapport 2005-13, Svenskt Vatten (Report, Swedish Water and Wastewater Association), Stockholm. pdf: www.svenkstvatten.se

Olofsson, A. (2004) *Juridiska förutsättning för uthålliga avloppssystem* (Legal prerequisites for alternative urban water systems, in Swedish). Urban Water Report 2004:2, Chalmers University of Technology, Göteborg. pdf: www.urbanwater.org

Olofsson, B., Tideström, H. and Willert, J. (2001) *Riskidentifiering av urbana VA-system* (Risk identification in Urban Water systems, in Swedish). Urban Water report 2001:2, Chalmers University of Technology, Göteborg. pdf: www.urbanwater.org

Ostrom, E. (1990) *Governing the Commons – The Evolution of Institutions for Collective Actions*, Cambridge University Press, Cambridge.

Ottoson, J. (2003) Hygiene aspects of greywater and greywater reuse. Licentiate Thesis, Royal Institute of Technology, Stockholm, Sweden.

Ottoson, J. and Stenström, T. A. (2003a) Faecal contamination of greywater and associated microbial risks. *Water Research* **37**(3), 645-655.

Ottoson, J. and Stenström, T. A. (2003b) Growth and reduction of microorganisms in sediments collected from a greywater treatment system. *Letters of Applied Microbiology* **36**(3), 168-72.

Ottoson, J. (2005) *Comparative analysis of pathogen occurrence in wastewater.* Ph.D. Thesis. Department of Land and Water Resource Engineering, Royal Institute of Technology (KTH), Stockholm, Sweden. pdf: www.urbanwater.org

Oxford English Dictionary Online (2004) 6 Sep. 2004. "Barrier, n." http://dictionary.oed.com/

Palmquist[1], H. and Jönsson, H. (2003) Urine, faeces, greywater, and biodegradable solid waste as potential fertilisers. *Proceedings of the 2nd Ecosan International Symposium of Ecological Sanitation*, 7th-11th April, Lübeck, Germany, pp. 587-594. Pdf: www.gtz.de/ecosan/download/ecosan-Symposium-Luebeck-session-f.pdf

Palmquist, H. (2004a) Substance Flow Analysis of Hazardous Substances in a Swedish Municipal Wastewater System. *Vatten* **60**(4), 251-260.

Palmquist, H. (2004b) *Hazardous substances in wastewater management.* Ph.D. Thesis. Dept. of Environmental Engineering. Luleå University of Technology, Luleå, Sweden. pdf: www.urbanwater.org

Palmquist, H. and Hanæus, J. (2005) Hazardous substances in separately collected grey- and blackwater from ordinary Swedish households. *Science of the Total Environment* **348**, 151-163.

---

[1] Now Almqvist

Payment, P. (1998) Distribution system impacts on microbial diseases. *Water Supply* **16**(3/4), 113-119.

Persson, F., Heinicke, G., Uhl, W, Hedberg, T. and Hermansson, M. (2005a) Biofiltration of surface water for removal of biodegradable organic matter and subsequent biofilm formation – a comparison of filter media. Subm. to *Env. Techn.*

Persson F., Långmark, J., Heinicke, G., Hedberg, T., Tobiason, J., Stenström, T.A. and Hermansson, M. (2005b) Characterisation of the behaviour of particles in biofilters for pre-treatment of drinking water. *Wat. Res.* **39**(16), 3791-3800.

Persson, F., Heinicke, G., Hedberg, T., Hermansson, M. and Uhl, W. (2005c) Removal of geosmin and MIB by biofiltration – an investigation discriminating between adsorption and biodegradation. Submitted to *Environ. Technol.*

Persson, F., Heinicke, G., Uhl, W, Huber, S., Hedberg, T. and Hermansson, M. (2005d) Nanofiltration of surface water with biofiltration or rapid media filtration as pre-treatment: Characterisation of membrane fouling. Submitted to *J. of Membrane Sci.*

Qdais, H.A. and Moussa, H. (2004) Removal of heavy metals from wastewater by membrane processes: a comparative study. *Desalination* **164**, 105-110.

Regli, S., Rose, J.B., Haas, C.N. and Gerba, C.P. (1991) Modeling the risk from *Giardia* and viruses in drinking water. *J. Amerc. Wat. Wks. Assoc.* **83**(11):76-84.

Richert-Stintzing A., Johansson, M. and Kvarnström, E. (2005) *Regionala aspekter av återföring av humanurin och andra avloppsfraktioner –fallstudie från Kullön i Vaxholm.* (Preliminary title, in Swedish).

Rodenburg, C., de Groot, H., Dalhuisen, J. and Nijkamp, P. (2001) *Water Management in Amsterdam. Strategic Report.* Economic and Social Institute, Vrije University, Amsterdam.

Roy, R. (2003) *Cost engineering: Why, what and how?* Decision Engineering Report Series. Edited by Roy, R. and Kerr, C. ISBN 1-861940-96-3. Cranfield University, Cranfield, UK.

Rulkens, W.H. (2004). Sustainable sludge management - what are the challenges for the future? *Water Science and Technology* **49**(10), 11-19.

Rydhagen, B. (2003) *Boendeaspekter på resurseffektiv hantering av svartvatten och organiskt hushållsavfall.* (User aspects on resource efficient handling of blackwater and organic kitchen waste, in Swedish) VA-Forsk rapport 2003-02, Svenskt Vatten (Report, Swedish Water and Wastewater Association), Stockholm. pdf: www.svenkstvatten.se

Segnestam, L. (2003) *Indicators of Environment and Sustainable Development. Theories and Practical Experience.* The World Bank: Environment Department Papers. Paper No. 89.

SEPA (1996) *Generella riktvärden för förorenad mark* (In Swedish). Rapport 4638, Naturvårdsverket (Report, Swedish Environmental Protection Agency), Stockholm. pdf: www.naturvardsverket.se

SEPA (1999) *Förorenade områden – Metodik för inventering* (In Swedish). Rapport 4918, Naturvårdsverket (Report, Swedish Environmental Protection Agency), Stockholm. pdf: www.naturvardsverket.se

SEPA (2000) *Bedömningsgrunder för miljökvalitet: Sjöar och Vattendrag* (In Swedish). Rapport 4913, Naturvårdsverket (Report, Swedish Environmental Protection Agency), Stockholm. pdf: www.naturvardsverket.se

SEPA (2002) *Metaller i stad och land.* (In Swedish). Rapport 5184, Naturvårdsverket (Report, Swedish Environmental Protection Agency), Stockholm. pdf: www.naturvardsverket.se

SEPA (2004) *Aktionsplan för återföring av fosfor ur avlopp* (In Swedish). Rappoï. ᴗᴗᴗ ᴧᴗ ᴗ ᴧᴗ Naturvårdsverket (Report, Swedish Environmental Protection Agency), Stockholm. pdf: www.naturvardsverket.se

Siebert, H. (1987) *Economics of the environment: theory and policy*, 2nd ed., Springer-Verlag, Berlin & New York.

SIWI (2004) *Stockholm Water Front – a forum for global water issues*. No. 3, September 2004. Stockholm International Water Institute (SIWI).

SMPA (2004) *Miljöpåverkan från läkemedel samt kosmetiska och hygieniska produkter* (Environmental effects of pharmaceuticals and cosmetic and sanitary products, In Swedish with English summary). Report, Swedish Medical Products Agency. pdf: www.mpa.se

Söderberg, H. (2002) Multicriteria through stakeholder dialogue – evaluation of a test case *Proceedings International Sustainable Development Research Conference*, Manchester, UK, April 8-9, 2002. pp. 446-454.

Söderberg, H. and Åberg, H. (2002) Assessing sustainable urban water systems from a socio-technical perspective – The case of Hammarby Sjöstad. *Water Science and Technology: Water Supply* 2(4), 203-210.

Söderberg, H., Etnier, C and Hellström, D. (2002) *Att väga samman det ojämförbara – betydelsen av beslutsunderlag, syntesstrategi och deltagande* (To compare the incomparable, in Swedish with extended English summary), Urban Water report 2002:1, Chalmers University of Technology, Göteborg. pdf: www.urbanwater.org

Söderberg, H. and Kärrman, E. (eds.) (2003) *Methodologies for Integration of Knowledge Areas. The case of Sustainable Urban Water Management.* Report no. 2003:15, Chalmers Architecture Publication Series, ISSN 1650-6340. Department of Built Environment and Sustainable Development Göteborg, Chalmers University of Technology, Göteborg.

Söderberg, H., Kain J-H., Åberg, H., van Moeffaert, D. and Kärrman, E. (2004) Evaluating NAIADE with respect to stakeholder participation. *Proceedings of the 4th International Conf. on Decision-Making in Urban & Civil Engineering*, Porto, Portugal.

Söderberg, H. and Kain, J-H. (2006) Assessment of Sustainable Waste Management Alternatives: How to Support Complex Knowledge Management. *Journal of Environmental Planning and Management* 49(1), 21-39.

Stark, S. (2005) *Phosphorous release and recovery from treated sewage sludge*. Ph.D. thesis. Dept. of Land and Water Resource Engineering, Royal Institute of Technology (KTH), Stockholm, Sweden. pdf: www.urbanwater.org

Stockholm Vatten (1999) *Föroreningsbelastning till sjön Trekanten. Utvärdering av beräkningsmodell för dagvatten* (Pollution load to Lake Trekanten. Evalution of a model for estimation of stormwater pollution, in Swedish). Rapport 44/99, Stockholm Vatten (Report, Stockholm Water Co.), Stockholm.

Storbjörk, S. and Söderberg, H. (2003) *Plötsligt händer det – Institutionella förutsättningar för uthålliga VA-system* (Suddenly it happens – institutional prerequisites for sustainable urban water systems, in Swedish with English summary), Urban Water report 2003:1, Chalmers University of Technology, Göteborg. pdf: www.urbanwater.org

Stradspan (2005) *Strategic Decision Support 2005.* www.btinternet.com/~stradspan/

Sugden, R. and Williams, A. (1978) *The principles of practical cost-benefit analysis.* Oxford University Press, Oxford, UK.

Svärd, Å. and Jansen, J.L.C. (2003) *Svenska biogasanläggningar – erfarenhets-sammanställning och rapporteringssystem* (Biogas plants in Sweden – compilation of experience and reporting system, in Swedish with English summary). VA-Forsk rapport 2003-14, Svenskt Vatten (Report, Swedish Water and Wastewater Association), Stockholm. pdf: www.svenkstvatten.se

Svensson, G., Malmqvist, P.-A. and Ahlman, S. (2001) Strategies for management of polluted storm water from an urban highway in Göteborg, Sweden. In: Marsalek, J., Watt, W.E., Zeman, E. and Sieker, H. (eds.). *Advances in Urban Stormwater and Agricultural Runoff Source Controls*, pp. 57-67, Kluwer Academic Publishers, Dordrecht, The Netherlands.

Thomasson, A., Mattisson, O. and Ramberg, U. (2005) *Uthålliga VA-system – internationella erfarenheter av organisationsformer och drivkrafter i en VA-sektor i utveckling* (Sustainable water and sewerage systems, in Swedish), VA-Forsk rapport 2005-2, Svenskt Vatten (Report, Swedish Water and Wastewater Association), Stockholm. pdf: www.svenkstvatten.se

Tidåker, P., Kärrman, E., Baky, A., Jönsson, H. (2005) Wastewater Management Integrated with Farming – An Environmental Systems Analysis of a Swedish County Town. Accepted for publication in *Resources, Conservation and Recycling*.

UNCED (1993) *The Earth Summit*. United Nations Conference on Environment and Development. Graham and Trotman Limited, London.

Uppsala municipality (2000) *Slamstrategi* (Strategy for sewage sludge, in Swedish), Uppsala municipality (unpublished)

Uppsala municipality (2002) *Översiktsplan för Uppsala stad* (General city plan of Uppsala, in Swedish), Uppsala municipality (unpublished)

van der Voet, E., Nikolic, I., Huppes, G. and Kleijn, R. (2004) Integrated systems analysis of persistent polar pollutants in the water cycle. *Water Science and Technology* **50**(5), 243-251.

van Moeffaert, D. (2003a) Multi-Criteria Decision Aid: Options and Experiences. In Söderberg, H. and E. Kärrman, (Eds.). *Methodologies for Integration of Knowledge Areas: The Case of Sustainable Urban Water Management*. Report 2003:15, Chalmers Architecture Publication Series, ISSN 1650-6340. Department of Built Environment and Sustainable Development Göteborg: Chalmers University of Technology, Göteborg.

van Moeffaert, D. (2003b) *Multi-criteria decision aid in sustainable urban water management*. Report 2002:26, Dept. of Industrial Ecology, Royal Institute of Technology (KTH), Stockholm, Sweden.

VASTRA (2005) Towards more integrated resource management – Experience from a Swedish Catchment. Special issue of *Ambio*, November 2005.

Vinnerås, B. and Jönsson, H. (2003) *Chemical disinfection of wastewater products* (Kemisk hygienisering av avloppsprodukter; in English, Swedish summary). VA-Forsk rapport 2003-24, Svenskt Vatten (Report, Swedish Water and Wastewater Association), Stockholm. pdf: www.svenkstvatten.se

Vinnerås, B., Jönssson, H. and Albihn, A. (2004) Evaluating two secondary treatment alternatives for sanitising separately dry collected faecal matter. *Proceedings from the IWA Biannual World Water Congress*, Sept 20-23, Marrakech, Morocco. CD-ROM.

Visvanathan, C. and Roy, P.K. (1997) Potential of nanofiltration for phosphate removal from wastewater. *Environmental Technology* **18**(5), 551-556.

Wallén, E. (1999) *Livscykelanalys av dricksvatten - en studie av ett vattenverk i Göteborg (LCA analysis of drinking water – A study of a waterworks in Göteborg, in Swedish)*. M.Sc. Thesis, Dept. of Technical Environmental Planning, Chalmers University of Technology, Göteborg.

Westrell, T., Bergstedt, O., Heinicke, G. and Kärrman, E. (2002) A systems analysis comparing drinking water systems – central physical-chemical treatment and local membrane filtration. *Wat. Sci. Tech.: Water Supply* **2**(2), 11-18.

Westrell, T., Bergstedt, O., Stenström, T.A. and Ashbolt, N.J. (2003) A theoretical approach to assess microbial risks due to failures in drinking water treatment. *International Journal of Environmental Health Research* **13**(2), 181-197.

Westrell, T. (2004) *Microbial risk assessment and its implications for risk management in urban water systems.* Ph.D. Thesis no. 304, Department of Water and Environmental Studies, Linköping Studies in Arts and Science, Linköping University, Linköping, Sweden. pdf: www.urbanwater.org

Westrell, T., Schönning, C., Stenström, T.A. and Ashbolt, N.J. (2004) QMRA (quantitative microbial risk assessment) and HACCP (hazard analysis and critical control points) for management of pathogens in wastewater and sewage sludge treatment and reuse. *Wat. Sci. Tech.* **50**(2), 23-30.

Westrell, T., Andersson, Y. and Stenström, T.A. (2006) Drinking water consumption patterns in Sweden. *J. Water and Health* (in print).

WHO and UNICEF (2000) *Global Water Supply and Sanitation Assessment 2000 Report.* WHO, Geneva, Switzerland. ISBN 9241562021.

WHO (2003) Faecal Pollution and Water Quality. Chapter 4 in: *Guidelines for Safe Recreational Water Environments. Vol. 1: Coastal and Fresh Waters.* World Health Organization, Geneva, Switzerland.

WHO (2004) *Guidelines for Drinking-water Quality,* 3rd ed., Volume 1: Recommendations. Geneva, Switzerland.

Wilsenach, J.A. and van Loosdrecht, M.C.M. (2003) Impact of separate urine collection on wastewater treatment systems. *Water Science and Technology* **48**(1), 103–110.

Worldwatch Institute (2004) *The State of the World 2004. Special Focus: The Consumer Society.* W.W. Norton & Company Inc., New York.

Åberg, H. (2000) *Sustainable waste management in households – from international policy to everyday practice. Experiences from two Swedish field studies.* Ph.D. Thesis, Göteborg Studies in Educational Sciences, no. 150, Göteborg University, Göteborg, Sweden.

Åberg, H. and Söderberg, H. (2003) Sustainable development and participation – why and when, by whom and how? In: Söderberg, H. and Kärrman, E. (eds.) *Methodologies for Integration of Knowledge Areas. The case of Sustainable Urban Water Management.* Report 2003:15, Chalmers Architecture Publication Series, ISSN 1650-6340. Department of Built Environment and Sustainable Development, Chalmers University of Technology, Göteborg, Sweden.

Åberg, H. (2004) *Boendeperspektiv på hushållsavfall och på system för insamling och behandling i Västra Hamnen, Malmö.* (Residents' perspectives on kitchen waste and systems for waste management in Västra Hamnen, Malmö, in Swedish) Research report no. 37, Dept of Home Economics, Göteborg University, Göteborg.

Index

Lightning Source UK Ltd.
Milton Keynes UK
14 January 2010

148583UK00001B/69/A

9 781843 391050